Clinical Pocket Manual™

Fluids and Electrolytes

NURSING86 BOOKS™
SPRINGHOUSE CORPORATION
SPRINGHOUSE, PENNSYLVANIA

Clinical Pocket Manual™ Series

PROGRAM DIRECTOR
Jean Robinson

CLINICAL DIRECTOR
Barbara McVan, RN

ART DIRECTOR
John Hubbard

EDITORIAL MANAGER
Susan R. Williams

EDITORS
Lisa Z. Cohen
Kathy E. Goldberg
Virginia P. Peck

CLINICAL EDITORS
Joan E. Mason, RN, EdM
Diane Schweisguth, RN, BSN

COPY SUPERVISOR
David R. Moreau

DESIGNER
Maria Errico

PRODUCTION COORDINATOR
Susan Powell-Mishler

The clinical procedures described and recommended in this publication are based on research and consultation with medical and nursing authorities. To the best of our knowledge, these procedures reflect currently accepted clinical practice; nevertheless, they can't be considered absolute and universal recommendations. For individual application, treatment recommendations must be considered in light of the patient's clinical condition, and before administration of new or infrequently used drugs, in light of latest package-insert information. The authors and the publisher disclaim responsibility for any adverse effects resulting directly or indirectly from the suggested procedures, from any undetected errors, or from the reader's misunderstanding of the text.

Material in this book was adapted from the following series: Nurse's Reference Library, Nursing photobook, New Nursing Skillbook, NursingNow, and Nurse's Clinical Library.

Amended reprint, 1986   CPM3-020586

© 1985 by Springhouse Corporation
1111 Bethlehem Pike, Springhouse, Pa. 19477. All rights reserved. Reproduction in whole or part by any means whatsoever without written permission of the publisher is prohibited by law. Printed in the United States of America.

Library of Congress Cataloging-in-Publication Data

Main entry under title:

Fluids and electrolytes.

(Clinical pocket manual)
"Nursing85 books."
Includes index.
1. Water-electrolyte imbalances—Handbooks, manuals, etc. 2. Water-electrolyte imbalances—Nursing—Handbooks, manuals, etc. I. Springhouse Corporation. II. Series. [DNLM: 1. Body Fluids—handbooks. 2. Body Fluids—nurses' instruction. 3. Water-Electrolyte Balance—handbooks. 4. Water-Electrolyte Balance—nurses' instruction. 5. Water-Electrolyte Imbalance—handbooks. 6. Water-Electrolyte Imbalance—nurses' instruction. WD 200.1 F646]
RC630.F592 1985   616.3'9   85-17254
ISBN 0-87434-003-9

TABLE OF CONTENTS

1 Body Fluids: Normal and Abnormal Values
Homeostasis 1-9; Blood 10-29; Pleural Fluid 30-31; Cerebrospinal Fluid 32-35; Peritoneal Fluid 36-37; Synovial Fluid 38-40; Urine 41-54

55 Electrolytes: Normal and Abnormal Values
General Information 55-59; Sodium 60-68; Potassium 69-80; Calcium 81-88; Magnesium 89-92; Other Electrolytes 93-98

99 Fluid and Electrolyte Imbalances
Third-Space Shift 99; Edema 100-102; Overhydration 103-104; Dehydration 105-111; Acid-Base Imbalances 112-116

117 Special Problems
Diabetes Mellitus 117-127; Diabetes Insipidus 128-131; Other Endocrine Disorders 132-135; Renal Failure 136-148; Hypovolemic Shock 149-153; Cardiac Disorders 154-158; Surgery 159-167; Burns 168-173; Pancreatitis 174

175 I.V. Hyperalimentation
Solutions 175-179; Complications 180-183

Nursing86 Books™

CLINICAL POCKET MANUAL™ SERIES
Diagnostic Tests
Emergency Care
Fluids and Electrolytes
Signs and Symptoms
Cardiovascular Care
Respiratory Care
Critical Care
Neurologic Care
Surgical Care

NURSING NOW™ SERIES
Shock
Hypertension
Drug Interactions
Cardiac Crises
Respiratory Emergencies
Pain

NURSE'S CLINICAL LIBRARY™
Cardiovascular Disorders
Respiratory Disorders
Endocrine Disorders
Neurologic Disorders
Renal and Urologic Disorders
Gastrointestinal Disorders
Neoplastic Disorders
Immune Disorders

NURSING PHOTOBOOK™ SERIES
Providing Respiratory Care
Managing I.V. Therapy
Dealing with Emergencies
Giving Medications
Assessing Your Patients
Using Monitors
Providing Early Mobility
Giving Cardiac Care
Performing GI Procedures
Implementing Urologic Procedures
Controlling Infection
Ensuring Intensive Care
Coping with Neurologic Disorders
Caring for Surgical Patients
Working with Orthopedic Patients
Nursing Pediatric Patients
Helping Geriatric Patients
Attending Ob/Gyn Patients
Aiding Ambulatory Patients
Carrying Out Special Procedures

NURSE'S REFERENCE LIBRARY®
Diseases
Diagnostics
Drugs
Assessment
Procedures
Definitions
Practices
Emergencies
Signs and Symptoms

NURSE REVIEW™ SERIES
Cardiac Problems
Respiratory Problems
Gastrointestinal Problems
Neurologic Problems
Vascular Problems

Nursing86 DRUG HANDBOOK™

HOMEOSTASIS

The Ins and Outs

In a healthy person, the fluids he ingests balance the fluids he excretes (see illustration below). Water loss via the skin and lungs will increase in a hot, dry environment or with increased respiratory rate, fever, or skin injury, such as burns. Water loss via the kidneys varies largely with the amount of solute excreted and with the level of antidiuretic hormone, which controls the kidneys' reabsorption of water.

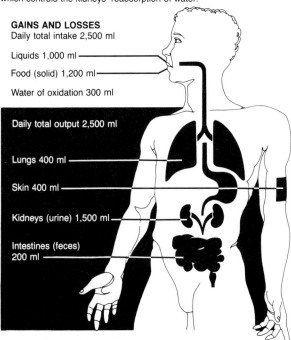

GAINS AND LOSSES
Daily total intake 2,500 ml

Liquids 1,000 ml

Food (solid) 1,200 ml

Water of oxidation 300 ml

Daily total output 2,500 ml

Lungs 400 ml

Skin 400 ml

Kidneys (urine) 1,500 ml

Intestines (feces) 200 ml

How to Determine Your Patient's Fluid Balance

To correctly determine your patient's fluid balance, you must maintain an accurate intake and output record. Every 24 hours, add all items on the record's intake portion. Then, add all items on the output portion. To find the 24-hour body fluid balance, combine the negative output sum and the positive intake sum.

Check this balance against the patient's daily weight, because his fluid loss or retention will be rapidly reflected in a weight change. A positive intake and output balance of 1 liter accounts for 2.2 lb (1 kg) of weight gain. A negative balance of the same amount accounts for 2.2 lb of weight loss.

24-HOUR INTAKE AND OUTPUT RECORD (in ml)

	11 to 7	7 to 3	3 to 11
ORAL INTAKE (Include tube feedings.)			
None (NG tube in place)	—	—	—
8-hour oral intake totals	0	0	0
PARENTERAL FLUID INTAKE (Blood, plasma, TPN, fat emulsions, I.V. solutions. Document time started.)			
1,000 ml D_5W with 0.45% NaCl at 11 p.m.	600	400	
1,000 ml D_5W at 10 a.m.		800	200
1,000 ml D_5W at 4 p.m.			900
8-hour parenteral fluid intake totals	600	1,200	1,100
ALL OTHER INTAKE (For example, irrigation fluid.)			
NG tube irrigation	120	120	120
8-hour all other intake totals	120	120	120
8-hour total intake	720 +	1,320 +	1,220
Total 24-hour intake			= 3,260

Continued

HOMEOSTASIS

How to Determine Your Patient's Fluid Balance
Continued

24-HOUR INTAKE AND OUTPUT RECORD (in ml)

	11 to 7	7 to 3	3 to 11
URINE OUTPUT (Urethral catheter-C; Ureteral-U.)			
Describe urine (if significant).			
Dark yellow, concentrated	600	1,000	800
8-hour urine output totals	600	1,000	800
ALL OTHER OUTPUT (Emesis, gastric drainage, wound drainage, blood, bile, liquid stool)			
NG tube drainage	200	150	120
Diaphoresis (Moderate-M; Profuse-P)	P	M	M
8-hour all other output totals	200	150	120
8-hour total output	800 +	1,150 +	920
Total 24-hour output			2,870

Weighing Your Patient Accurately

When obtaining your patient's daily weight, observe the following guidelines:
- Use the same scale each day.
- Weigh the patient at the same time each day (preferably before he eats breakfast).
- Make sure the patient's wearing the same (or similar) clothing each time you weigh him.
- Remove all items that would add to the patient's own weight. If you can't remove an item, weigh its equivalent separately. Then, subtract its weight from the patient's total weight.
- Correlate the patient's weight with a 24-hour tally of his intake and output record.

Some Laboratory Tests for Evaluating Fluid Status

LAB TEST	NORMAL VALUE	SIGNIFICANCE
Serum osmolality Measures particles exerting osmotic pull per unit of water; reflects total body hydration	280 to 294 mOsm/kg	• Increases in dehydration • Decreases with water overload
Blood urea nitrogen (BUN) Reflects difference between rate of urea synthesis and its excretion by the kidneys	10 to 20 mg/100 ml	• Increases with decreased renal blood flow or urine production, dehydration, some neoplasms, and certain antibiotics • Decreases in pregnancy, overhydration, severe liver disease, and malnutrition
Hematocrit Measures portion of blood volume occupied by RBCs	Female: 37 to 47 ml/100 ml Male: 40 to 54 ml/100 ml	• Increases in dehydration • Decreases with low RBC count or with normal hemoglobin and water overload
Creatinine (serum) Measures products of muscle metabolism	0.5 to 1.5 mg/100 ml	• Elevated when 50% or more of the nephrons are destroyed

Continued

Some Laboratory Tests for Evaluating Fluid Status
Continued

LAB TEST	NORMAL VALUE	SIGNIFICANCE
Urine osmolality Measures number of particles per unit of water in urine	50 to 1,200 mOsm/liter	• Reflects changes in urine contents more accurately than specific gravity but depends on the prior state of hydration
Urine specific gravity	1.010 to 1.030	• Increases with any condition causing hypoperfusion of kidneys, which may lead to oliguria • Decreases when renal tubules cannot reabsorb water and concentrate urine, as in early pyelonephritis
Urine pH	4.6 to 8.0 (Average 6.0)	• Increases in metabolic and respiratory alkalosis; in the presence of magnesium ammonium phosphate stones; with certain urea-splitting infections • Decreases in the presence of uric acid stones and acidosis • In renal acidosis, pH may be normal or slightly more acidic only when the plasma HCO_3^- is very low

Homeostasis: Many Organs Take Part

Many organs control the body's homeostasis. The *pituitary gland* controls secretion of antidiuretic hormone, renal filtration and plasma flow, and adrenal function. The *parathyroids* maintain the level of ionized calcium in the blood and regulate the kidneys' conservation of magnesium. The *lungs* govern the exhalation or retention of carbon dioxide, which influences the body's acid-base balance. The *heart* and *vascular system* nourish the *kidneys*, which, through filtration, reabsorption, and excretion, control the necessary balance of fluids and electrolytes. The *adrenal glands* secrete aldosterone, which affects the retention or excretion of sodium, potassium, and water.

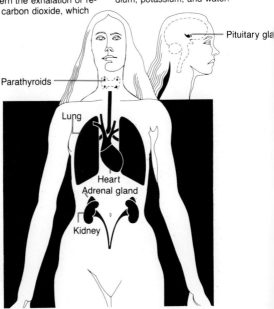

Hormones for Homeostasis

Two main hormones affect the balance of water and sodium in the body. The hypothalamus releases antidiuretic hormone (ADH). ADH acts on the distal renal tubules and the collecting duct of the kidney to reabsorb or excrete water.

Midbrain volume receptors, sensitive to the serum sodium load, stimulate the adrenal gland to release aldosterone. Aldosterone controls the reabsorption and excretion of sodium.

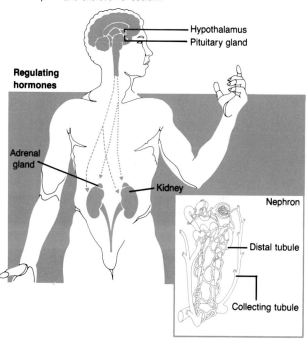

HOMEOSTASIS

Hormonal Secretion Sites

Powerful, complex chemicals, hormones circulate through the bloodstream to stimulate or inhibit the activity of target glands or organs. In coordination with the nervous system, these glands secrete hormones that maintain homeostasis. Some of these hormones are listed below.

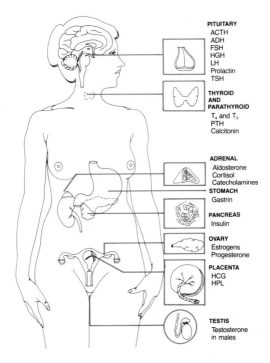

PITUITARY
ACTH
ADH
FSH
HGH
LH
Prolactin
TSH

THYROID AND PARATHYROID
T_4 and T_3
PTH
Calcitonin

ADRENAL
Aldosterone
Cortisol
Catecholamines

STOMACH
Gastrin

PANCREAS
Insulin

OVARY
Estrogens
Progesterone

PLACENTA
HCG
HPL

TESTIS
Testosterone in males

Understanding Osmolarity

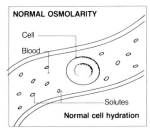

- 285 to 300 mOsm/liter

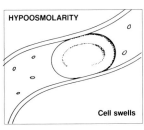

- 260 to 275 mOsm/liter (restlessness, confusion)

- 240 to 259 mOsm/liter (muscle aches, twitches)

- < 240 mOsm/liter (seizures, water intoxication)

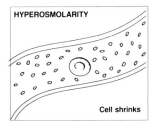

- 350 to 375 mOsm/liter (restlessness, hyperirritability)

- 376 to 400 mOsm/liter (nystagmus, tremors, progressive lethargy)

- > 400 mOsm/liter (coma)

Interpreting a Complete Blood Count (CBC)

The CBC gives a fairly complete picture of all the blood's formed elements and includes determinations of hemoglobin concentration, hematocrit, red and white counts, differential white cell count, and stained red cell examination. Besides pointing the way toward further studies, CBC data are valuable in themselves.

For example, they can detect anemias, determine their severity, and compare the status of specific blood elements.

WHITE BLOOD CELLS (WBCs)

Normal values
4,100 to 10,900/μl

Possible causes of abnormality
Above normal:
Infection (such as an abscess, meningitis, appendicitis, or tonsillitis), leukemia, tissue necrosis (caused, for example, by burns, myocardial infarction, or gangrene), stress, allergy
Below normal:
Viral infection (such as influenza, measles, infectious hepatitis, mononucleosis, or rubella), typhoid fever, bone marrow depression (caused, for example, by administration of antineoplastic drugs, ingestion of mercury or other heavy metals, or exposure to benzene or arsenic)

WBC DIFFERENTIAL

LYMPHOCYTES
Normal values (percentage of total WBC count)
16.2% to 43%

Possible causes of abnormality
Above normal:
- Infections: pertussis, brucellosis, syphilis, tuberculosis, hepatitis, infectious mononucleosis, mumps, German measles, cytomegalovirus
- Other: thyrotoxicosis, hypoadrenalism, ulcerative colitis, immune diseases, lymphocytic leukemia

Below normal:
- Severe debilitating illness, such as CHF, renal failure, advanced TB
- Defective lymphatic circulation, high levels of adrenal corticosteroids, immunodeficiency due to immunosuppressives

Continued

Interpreting a Complete Blood Count (CBC)
Continued

WBC DIFFERENTIAL *Continued*

NEUTROPHILS
Normal values (percentage of total WBC count)
47.6% to 76.8%

Possible causes of abnormality
Above normal:
- Infections: osteomyelitis, otitis media, salpingitis, septicemia, gonorrhea, endocarditis, smallpox, chicken pox, herpes, Rocky Mountain spotted fever
- Ischemic necrosis due to myocardial infarction, burns, carcinoma
- Metabolic disorders: diabetic acidosis, eclampsia, uremia, thyrotoxicosis
- Stress response due to acute hemorrhage, surgery, excessive exercise, emotional distress, third trimester of pregnancy, childbirth
- Inflammatory disease: rheumatic fever, rheumatoid arthritis, acute gout, vasculitis, myositis

Below normal:
- Bone marrow depression due to radiation or cytotoxic drugs
- Infections: typhoid, tularemia, brucellosis, hepatitis, influenza, measles, mumps, rubella, infectious mononucleosis
- Hypersplenism: hepatic disease and storage diseases
- Collagen vascular disease
- Deficiency of folic acid or vitamin B_{12}

EOSINOPHILS
Normal values (percentage of total WBC count)
0.3% to 7%

Possible causes of abnormality
Above normal:
- Allergic disorders: asthma, hay fever, food or drug sensitivity, serum sickness, angioneurotic edema
- Parasitic infections: trichinosis, hookworm, roundworm, amebiasis
- Skin diseases: eczema, pemphigus, psoriasis, dermatitis, herpes
- Neoplastic diseases: chronic myelocytic leukemia, Hodgkin's disease, metastasis and necrosis of solid tumors
- Miscellaneous: collagen vascular disease, adrenocortical hypofunction, ulcerative colitis, polyarteritis nodosa, postsplenectomy, pernicious anemia, scarlet fever, excessive exercise

Below normal:
- Stress response due to trauma, shock, burns, surgery, mental distress
- Cushing's syndrome

Continued

Interpreting a Complete Blood Count (CBC)
Continued

WBC DIFFERENTIAL *Continued*

MONOCYTES
Normal values (percentage of total WBC count)
0.6% to 9.6%

Possible causes of abnormality
Above normal:
- Infections: subacute bacterial endocarditis, tuberculosis, hepatitis, malaria, Rocky Mountain spotted fever
- Collagen vascular disease: systemic lupus erythematosus, rheumatoid arthritis
- Carcinomas
- Monocytic leukemia
- Lymphomas

BASOPHILS
Normal values (percentage of total WBC count)
0.3% to 2%

Possible causes of abnormality
Above normal:
Chronic myelocytic leukemia, polycythemia vera, some chronic hemolytic anemias, Hodgkin's disease, systemic mastocytosis, myxedema, ulcerative colitis, chronic hypersensitivity states, nephrosis

Below normal:
Hyperthyroidism, ovulation, pregnancy, stress

RED BLOOD CELLS (RBCs)

Normal values
Men: 4.5 to 6.2 million/μl
Women: 4.2 to 5.4 million/μl

Possible causes of abnormality
Above normal:
Dehydration, polycythemia
Below normal:
Hemorrhage, anemia from dietary deficiencies, sickle cell anemia, fluid overload

HEMOGLOBIN

Normal values
Men: 14 to 18 g/dl (after age 40, 12.4 to 14.9 g/dl)
Women: 12 to 16 g/dl (after age 40, 11.7 to 13.8 g/dl)

Possible causes of abnormality
Above normal:
Hemoconcentration (from polycythemia or dehydration)
Below normal:
Anemia, pregnancy, recent hemorrhage, fluid retention (causing hemodilution)

Continued

Interpreting a Complete Blood Count (CBC)
Continued

HEMATOCRIT

Normal values
Men: 42% to 54% of total blood volume
Women: 38% to 46% of total blood volume

Possible causes of abnormality
Above normal:
Dehydration, polycythemia
Below normal:
Anemia, acute hemorrhage, fluid overload

PLATELETS

Normal values
130,000 to 370,000/mm^3

Possible causes of abnormality
Above normal:
Hemorrhage, hemoconcentration, pregnancy, inflammatory or infectious disorders, splenectomy, polycythemia, iron deficiency anemia
Below normal:
Aplastic anemia, acute leukemia, thrombocytopenic purpura, administration of antineoplastic drugs, disseminated intravascular coagulation, infection

ERYTHROCYTE SEDIMENTATION RATE

Normal values
Men: 0 to 10 mm/hr
Women: 0 to 20 mm/hr

Possible causes of abnormality
Above normal:
Tissue destruction (either inflammatory or degenerative), pregnancy, menstruation, acute fever, cancer, infection, anemia
Below normal:
Liver disease, polycythemia, sickle cell anemia, blood hyperviscosity, low-serum protein

Normal Hematocrit Varies According to Age

NEWBORN	55% to 68%
1 WEEK	47% to 65%
1 MONTH	37% to 49%
3 MONTHS	30% to 36%
1 YEAR	29% to 41%
10 YEARS	36% to 40%
ADULT MALE	42% to 54%
ADULT FEMALE	38% to 46%

Guide to Whole Blood and Its Components

TYPE/ DESCRIPTION	INDICATIONS	CONTRAINDICATIONS
Whole blood Complete, unadulterated blood	• To restore an adequate volume of blood in hemorrhaging, trauma, or burn patients	• When the patient doesn't need volume increase and a specific component is available
Red blood cells (packed, frozen) Whole blood with 80% of the supernatant plasma removed	• To correct red blood cell deficiency and improve oxygen-carrying capacity of blood • To transfuse organ transplant patients or for patients with repeated febrile transfusion reactions, use frozen-thawed RBCs.	• When the patient's anemic from a deficiency of the hematopoietic nutrients; for example, iron, vitamin B_{12}, or folic acid • When the patient's asymptomatic, but you must raise his hematocrit level
White blood cells (leukocyte concentrate) Whole blood with all the red blood cells and 80% of the supernatant plasma removed	• To treat the patient who has life-threatening granulocytopenia from intensive chemotherapy, especially if his infections aren't responsive to antibiotics	• When the patient's health depends on the recovery of his bone marrow functions

CROSS MATCHING	ADMINISTRATION TECHNIQUES	SPECIAL CONSIDERATIONS
Necessary	• Administer by straight line set, Y-set, or microaggregate recipient set.	• Although whole blood is seldom transfused, components that are necessary—and frequently administered—are extracted from it. • Plasma protein fraction or albumin are given as volume expanders.
Necessary	• Administer by straight line set, Y-set, or microaggregate recipient set.	• Red blood cells have the same oxygen-carrying capacity as whole blood without the hazards of overload. Their use avoids the buildup of potassium and ammonia that sometimes occurs in the plasma of stored blood. Frozen-thawed RBCs are extremely expensive.
Must be ABO compatible (preferably human leukocyte group A antigen compatible)	• Administer by any straight line set. • Use with standard in-line blood filter. • Dosage: 1 unit daily until infection clears (usually within 5 days)	*Important:* WBC infusion *induces* a fever and can cause mild hypertension, severe chills, disorientation, and hallucinations. Try to control the patient's chills with antipyretics or blankets. Then treat the hypertension, if necessary.

Continued

Guide to Whole Blood and Its Components
Continued

TYPE/ DESCRIPTION	INDICATIONS	CONTRA- INDICATIONS
Plasma (fresh, fresh frozen) Uncoagulated plasma separated from whole blood	• To treat a clotting factor deficiency (when specific concentrates are unavailable or precise deficiency's unknown), hypovolemia, or a patient with a severe hepatic disease who has a limited synthesis of plasma coagulation factors • To prevent dilutional hypocoagulability	• When blood coagulation can be corrected with available specific therapy • When patient needs only albumin
Platelets Platelet sediment from platelet-rich plasma, resuspended in 30 to 50 ml of plasma	• To treat the patient with thrombocytopenia whose bleeding is caused by the following: decreased platelet production, increased platelet destruction, functionally abnormal platelets, or massive transfusions of stored blood (dilutional thrombocytopenia)	• When bleeding's unrelated to decreased number of platelets or abnormal function of platelets • When the patient's suffering from post-transfusion purpura or thrombotic thrombocytopenic purpura

CROSS MATCHING	ADMINISTRATION TECHNIQUES	SPECIAL CONSIDERATIONS
Unnecessary	• Administer as rapidly as possible by any straight line set.	• Normal saline solution not needed for Y-set, because the component contains no RBCs
Unnecessary (donor plasma and recipient's RBCs should be ABO compatible)	• Administer as rapidly as possible (uninterrupted) by syringe or component drip set only. • Don't use a standard in-line blood filter. Instead, use a nonwettable filter. • Dosage: 2 units per kg of body weight to raise the platelet count by at least 50,000/mm^3	• Usually administered when patient's platelet count drops below 10,000/mm^3 • For the patient with a history of side effects; administer antihistamines before platelet transfusion. Slower administration may be necessary to prevent overload. • Least hazardous when given fresh • Must be constantly agitated during storage

Continued

Guide to Whole Blood and Its Components
Continued

TYPE/ DESCRIPTION	INDICATIONS	CONTRA- INDICATIONS
Plasma protein fraction 5% solution of selected proteins from pooled plasma in a buffered, stabilized saline diluent	• To treat hypovolemic shock or hypoproteinemia • For initial treatment of infants in shock or children who are dehydrated or who have electrolyte deficiencies • May be used cautiously when the patient has congestive heart failure from added fluid and salt load or has renal or hepatic failure from added protein load	• When the patient has a clotting factor deficiency
Albumin 5% (buffered saline) **Albumin 25%** (salt-poor) Heat-treated, aqueous, chemically processed fraction of pooled plasma	• To treat shock from burns, trauma, surgery, or infections • To prevent marked hemoconcentration • To maintain appropriate electrolyte balance • To treat hypoproteinemia (with or without edema)	• When the patient has severe anemia or cardiac failure

BLOOD

CROSS MATCHING	ADMINISTRATION TECHNIQUES	SPECIAL CONSIDERATIONS
Unnecessary	• Administer by any straight line set at a rate and volume dependent on the patient's condition and response.	• Should not be mixed in the same line with protein-hydrolysates and alcohol solutions • Often given as a volume expander in place of whole blood while cross matching is being completed
Unnecessary	• Administer by provided set undiluted, or diluted with saline solution or 5% dextrose in water. • For a patient with normal blood volume or hypoproteinemia; administer slowly (1 ml/min) to prevent rapid plasma volume expansion. • For a patient in shock; administer as rapidly as possible.	• Can't transmit hepatitis because it's heat-treated at 140° F. (60° C.) for 10 hours • Often given as a volume expander in place of whole blood while cross matching is being completed

Understanding Routine Blood Studies

In this chart, you'll find the normal levels for the most commonly analyzed blood substances and possible reasons for abnormal levels. However, keep in mind that levels considered normal may vary somewhat from hospital to hospital depending on the specific testing methods used. Use the values from your hospital's laboratory manual.

ALBUMIN

Normal range
3.5 to 5 g/dl
Possible causes of abnormality
Above normal:
- Multiple myeloma

Below normal:
- Liver or renal disease
- Hodgkin's disease
- Peptic ulcer
- Malnutrition
- Plasma loss from burns
- Diarrhea
- Acute cholecystitis
- Collagen disease
- Hyperthyroidism

BILIRUBIN, TOTAL

Normal range
0.1 to 1 mg/dl
Possible causes of abnormality
Above normal:
- Liver damage
- Pernicious anemia
- Eclampsia
- Biliary obstruction

ALKALINE PHOSPHATASE

Normal range
Men: 90 to 239 units/liter
Women under age 45:
76 to 196 units/liter
Women over age 45:
87 to 250 units/liter
Possible causes of abnormality
Above normal:
- Skeletal disease (rickets or osteomalacia)
- Hyperparathyroidism
- Mononucleosis
- Liver disease

BLOOD UREA NITROGEN

Normal range
8 to 20 mg/dl
Possible causes of abnormality
Above normal:
- Mercury poisoning
- Obstructive uropathy
- Renal disease
- Excessive protein in diet
- Digestion of blood from gastrointestinal tract bleeding

Below normal:
- Severe hepatic damage
- Malnutrition
- Overhydration

Continued

Understanding Routine Blood Studies
Continued

TOTAL CARBON DIOXIDE CONTENT

Normal range
22 to 34 mEq/liter
Possible causes of abnormality
Above normal:
- Respiratory acidosis
- Intestinal obstruction
- Vomiting or continuous gastric drainage, causing metabolic alkalosis
- Primary aldosteronism

Below normal:
- Metabolic acidosis; for example, diabetic acidosis
- Renal failure
- Diarrhea or continuous intestinal drainage
- Anesthesia, causing change in respiratory pattern

CREATININE

Normal range
Men: 0.8 to 1.2 mg/dl
Women: 0.6 to 0.9 mg/dl
Possible causes of abnormality
Above normal:
- Destruction of more than 50% of the nephrons
- Acromegaly
- Gigantism

CALCIUM

Normal range
8.9 to 10.1 mg/dl
Possible causes of abnormality
Above normal:
- Hyperparathyroidism
- Multiple myeloma
- Excessive ingestion of milk or antacids containing calcium
- Multiple fractures
- Prolonged bed rest

Below normal:
- Hypoparathyroidism
- Malnutrition
- Calcium malabsorption
- Renal failure
- Administration of large volumes of citrated, stored blood
- Acute pancreatitis

CHOLESTEROL

Normal range
120 to 330 mg/dl
Possible causes of abnormality
Above normal:
- Nephrotic syndrome
- Bile duct blockage
- Hypothyroidism
- Pancreatitis
- Lipemia
- Excessive cholesterol in diet

Below normal:
- Malnutrition
- Pernicious and hemolytic anemia
- Hyperthyroidism

Continued

Understanding Routine Blood Studies
Continued

GLUCOSE (FASTING)

Normal range
80 to 120 mg/dl
Possible causes of abnormality
Above normal:
- Diabetes mellitus
- Pancreatitis
- Hepatic disease
- Anoxia
- Convulsive disorders
- Administration of excessive glucose I.V.

Below normal:
- Strenuous exercise
- Hyperinsulinism
- Insulinoma
- Hepatic insufficiency
- Addison's disease
- Hypoglycemia

POTASSIUM

Normal range
3.8 to 5.5 mEq/liter
Possible causes of abnormality
Above normal:
- Renal failure
- Addison's disease
- Blood specimen hemolysis
- Tissue breakdown or hemolysis from injury
- Excessive administration of potassium supplements

Below normal:
- Excessive use of diuretics
- Cushing's syndrome

LACTIC DEHYDROGENASE

Normal range
100 to 225 milliunits (mU) per ml
Possible causes of abnormality
Above normal:
- Pulmonary infarction
- Liver disease
- Shock
- Myocardial infarction
- Renal, brain, or skeletal muscle damage
- Pernicious, hemolytic, or sickle cell anemia
- Muscular dystrophy
- Blood specimen hemolysis

SODIUM

Normal range
135 to 145 mEq/liter
Possible causes of abnormality
Above normal:
- Dehydration
- Pyloric obstruction
- Administration of hypertonic saline solution

Below normal:
- Excessive use of diuretics
- Vomiting
- Diarrhea
- Diaphoresis
- Gastrointestinal suctioning
- Inadequate sodium intake
- Adrenal insufficiency

Continued

Understanding Routine Blood Studies
Continued

PROTEIN, TOTAL

Normal range
6.6 to 7.9 g/dl
Possible causes of abnormality
Above normal:
- Dehydration
- Diabetic acidosis
- Chronic infection
- Multiple myeloma
- Monocytic leukemia
- Shock

Below normal:
- Hemorrhage
- Malnutrition
- Hodgkin's disease
- Blood dyscrasia
- Eclampsia
- Hypertension
- Severe burns

SERUM GLUTAMIC-OXALO-ACETIC TRANSAMINASE

Normal range
8 to 20 units/liter

Possible causes of abnormality
Above normal:
- Hepatitis
- Cholecystitis
- Myocardial infarction
- Mononucleosis

URIC ACID

Normal range
Men: 4.3 to 8 mg/dl
Women: 2.3 to 6 mg/dl
Possible causes of abnormality
Above normal:
- Gout
- Leukemia
- Impaired renal function
- Toxemia of pregnancy

Below normal:
- Defective reabsorption of uric acid by kidney tubules, possibly from shock
- Acute hepatic atrophy

Special Consideration

Special considerations for venipunctures: Never draw a venous sample from an arm or leg that is being used for I.V. therapy, blood administration, or I.V. drug administration. Also, a venous sample should not be drawn from a site of infection.

If you use a blood pressure cuff as a tourniquet, inflate it to a range between the patient's systolic and diastolic pressures *to allow for venous distention without constricting arterial flow.*

Blood Gases and Electrolyte Definitions

Partial pressure	A measure of the force that a gas exerts on the fluid in which it is dissolved
Pao_2	Partial pressure of oxygen in arterial blood
$Paco_2$	Partial pressure of carbon dioxide in arterial blood
pH	A measure of acidity or alkalinity, or the concentration of free hydrogen ions in the blood
O_2CT	Oxygen content, or the volume of oxygen combined with hemoglobin in arterial blood
O_2 Sat	Oxygen saturation, a measure of the percentage of oxygen combined with hemoglobin to the total amount of oxygen with which hemoglobin could combine
Electrolytes	Substances that dissociate into ions when fused or in solution and thus conduct electricity
Cations	Positively charged ions
Anions	Negatively charged ions
Acidemia	Accumulation of acid or a loss of base that can result from metabolic or respiratory changes
Alkalemia	Accumulation of base or a loss of acid that can result from metabolic or respiratory changes

Respiratory Exchange Rate

This illustration shows the pattern of respiratory gas exchange within the human body. Remember that atmospheric pressure is 760 mm Hg; oxygen (O_2) in dry air, 21%, or 160 mm Hg partial pressure (PO_2); and nitrogen (N_2) in dry air, 79%, or 600 mm Hg partial pressure (PN_2). The boxed numbers in the illustration are especially important.

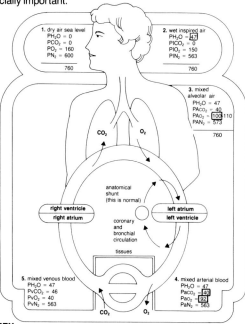

ART KEY
I = inspired   v = venous   A = alveolar   a = arterial

How to Interpret Laboratory Values

Oxygen tension:
The PaO_2 represents the oxygen saturation of arterial blood. You need to know the following values for PaO_2:
• 80 to 100 mm Hg, normal
• 70 mm Hg or below, hypoxemia
• 50 mm Hg or below, dangerous hypoxemia.
Assuming the patient has no lung disease, giving oxygen greatly increases PaO_2. Giving 100% oxygen (when room air is only 21% oxygen) can produce an arterial oxygen tension well above 600 mm Hg in a healthy person. So you must take oxygen administration into account when you evaluate the results of blood-gas analysis. Consider, for example, that a patient is receiving oxygen by nasal cannula at a flow rate of 7 liters/minute and his PaO_2 is 90 mm Hg. Normally, a flow rate of 7 liters/minute of oxygen results in a PaO_2 around 200 mm Hg. If the patient receiving this much oxygen has a PaO_2 of only 90 mm Hg, you can be sure he has some respiratory insufficiency. Without supplementary oxygen, this PaO_2 is in normal range.

Carbon dioxide tension:
The $PaCO_2$ is the carbon dioxide tension of arterial blood and is the single best measure of adequacy of alveolar ventilation. You need to know the following values for $PaCO_2$:
• 34 to 44 mm Hg, normal; ventilation is adequate
• above 46 mm Hg, hypercarbia; hypoventilation
• below 35 mm Hg, hypocarbia; hyperventilation

Continued

How to Interpret Laboratory Values
Continued

Acid-base balance:
The pH represents the acidity or alkalinity of the blood in terms of its concentration of free hydrogen ions (H^+). The greater their concentration, the more acid the solution and the *lower* the pH. The pH is affected by both metabolic and respiratory factors, as well as by chemical buffering systems in the body. Remember the following values for pH and what they mean:

- 6.80 to 7.00, severe, life-threatening acidemia; incompatible with life if untreated; immediate intervention required.

- 7.00 to 7.35, acidemia, producing acidosis

- 7.35 to 7.45, normal

- 7.45 to 7.70, alkalemia, producing alkalosis

- 7.70 to 7.80, severe, life-threatening alkalemia; immediate intervention required.

Norms for Vital Signs and Blood Gases in Children

The child's age, activity level, and physical and emotional status can affect his vital signs. In fact, even changing the time of day when you routinely measure his *body temperature, pulse rate, respirations,* and *blood pressure* can result in variations from the previous to the current determinations. Keep these considerations in mind when you assess your patient's vital signs.

	0 to 1 YEAR
Blood pressure	90 ± 25 systolic 61 ± 19 diastolic
Pulse	70 to 180
Temperature	99.6° F. (37.6° C.) rectal 98.6° F. (37° C.) oral 97.4° F. (36.3° C.) axillary
Respirations	30 to 40
Blood gases	PaO_2 85 to 100 $PaCO_2$ 35 to 45

BLOOD

BODY FLUIDS: NORMAL/ABNORMAL VALUES

Taking a patient's vital signs allows you to: determine the relative status of vital organs, including the heart, blood vessels, and lungs; monitor response to treatment; establish baseline measurements that can be compared with future readings; and determine the need for further diagnostic testing.

1 to 2 YEARS	2 to 3 YEARS	3 to 4 YEARS
96 ± 27 systolic 65 ± 27 diastolic	95 ± 24 systolic 61 ± 24 diastolic	99 ± 23 systolic 65 ± 19 diastolic
80 to 140	80 to 140	80 to 120
99.6° F. rectal 98.6° F. oral 97.4° F. axillary	99.6° F. rectal 98.6° F. oral 97.4° F. axillary	99.6° F. rectal 98.6° F. oral 97.4° F. axillary
28 to 32	28 to 32	24 to 28
PaO_2 85 to 100 $PaCO_2$ 34 to 45	PaO_2 85 to 100 $PaCO_2$ 35 to 45	PaO_2 85 to 100 $PaCO_2$ 35 to 45

Examining Pleural Fluid

Appearance of pleural fluid	Possible cause
Light, straw-colored	• Normal
Purulent	• Empyema
Blood-tinged	• Hemothorax • TB • Pulmonary infarction • Neoplastic disease • Accidental tissue damage from thoracentesis
Milky (chylothorax)	• Thoracic tumor or an inflammatory process • Traumatic rupture of thoracic duct • Cellular debris or cholesterol crystals

Test	Interpretation
Gram stain culture and sensitivity	• If positive, it may mean the early stages of bacterial infection. *Remember:* In the later stages of bacterial infection, the fluid may look grossly purulent with a positive Gram stain, yet cultures may be negative from antibiotic therapy.
Acid-fast stain and culture	• If positive, it may mean TB.
Red blood cells	• If count is about 10,000 per mm^3 and the specimen's pink or light red in color, it may indicate tissue damage.

Continued

PLEURAL FLUID

Test	Interpretation
Red blood cells *Continued*	• If count is above 100,000 per mm^3 and the specimen's grossly bloody, it suggests intrapleural malignancy, pulmonary infarction, TB, or chest trauma.
Leukocytes	• If count is above 1,000 per mm^3 or above 50% neutrophils, it may indicate inflammation.
Lymphocytes	• If count is over 50%, it may mean TB, lymphoma, or neoplasm.
Blood clots	• If present, they may mean neoplasm, TB, or infection.
Specific gravity	• Above 1.016 may mean neoplasm, TB, or infection. • Less than 1.104 may mean CHF.
Total protein	• Above 3 may mean CHF. • Less than 3 suggests neoplasm, TB, or infection.
Lactic dehydrogenase (LDH)	• Increased in cancer. • Decreased in heart failure.
Glucose	• If less than serum glucose, it may mean cancer or a bacterial infection.
Sediment	• If present, it may be cancerous cells, cellular debris, or cholesterol crystals.

CSF Findings

CSF is a unique distillate of blood that originates in the choroid plexuses of the brain and circulates in the subarachnoid space. It protects the brain and spinal cord from injury and transports products of neurosecretion, cellular biosynthesis, and cellular metabolism through the central nervous system (CNS). For laboratory analysis, CSF is obtained most commonly by lumbar puncture (usually between the third and fourth lumbar vertebrae) and, occasionally, by cisternal or ventricular puncture.

TEST	NORMAL
Pressure	50 to 180 mm H_2O
Appearance	Clear, colorless
Protein	15 to 45 mg/100 ml
Gamma globulin	3% to 12% of total protein

CEREBROSPINAL FLUID

A sample of CSF is frequently obtained during other neurologic tests, including myelography and pneumoencephalography. Infection at the puncture site contraindicates removal of CSF. In a patient with increased ICP, CSF should be removed with extreme caution because the rapid reduction in pressure that follows withdrawal of fluid can cause cerebellar tonsillar herniation and medullary compression.

ABNORMAL	IMPLICATIONS
Increase	Increased intracranial pressure due to hemorrhage, tumor, or edema caused by trauma
Decrease	Spinal subarachnoid obstruction above puncture site
Cloudy	Infection (elevated WBC count and protein, or many microorganisms)
Xanthochromic or bloody	Subarachnoid, intracerebral, or intraventricular hemorrhage; spinal cord obstruction; traumatic tap (usually noted only in initial specimen)
Brown, orange, or yellow	Elevated protein, RBC breakdown (blood present for at least 3 days)
Marked increase	Tumors, trauma, hemorrhage, diabetes mellitus, polyneuritis, blood in CSF
Marked decrease	Rapid CSF production
Increase	Demyelinating disease (such as multiple sclerosis), neurosyphilis, Guillain-Barré syndrome

Continued

CSF Findings
Continued

TEST	NORMAL
Glucose	50 to 80 mg/100 ml (⅔ of blood glucose)
Cell count	0 to 5 WBCs
	No RBCs
VDRL and other serologic tests	Nonreactive
Chloride	118 to 130 mEq/liter
Gram stain	No organisms

Post-Test Care

- Check if the patient must lie flat or if the head of his bed may be slightly elevated. In most cases, you will be instructed to *keep the patient lying flat for 8 hours after lumbar puncture.* Some doctors, however, allow a 30° elevation at the head of the bed. Remind the patient that although he must not raise his head, he can turn from side to side.
- Encourage the patient to drink fluids. Provide a flexible straw so he won't have to lift his head.
- Check the puncture site for redness, swelling, and drainage every 4 hours for the first 24 hours.

ABNORMAL	IMPLICATIONS
Increase	Systemic hyperglycemia
Decrease	Systemic hypoglycemia, bacterial or fungal infection, meningitis, mumps, postsubarachnoid hemorrhage
Increase	Active disease: meningitis, acute infection, onset of chronic illness, tumor, abscess, infarction, demyelinating disease
RBCs	Hemorrhage or traumatic tap
Positive	Neurosyphilis
Decrease	Infected meninges
Gram-positive or gram-negative organisms	Bacterial meningitis

Post-Test Care
Continued

- If CSF pressure is elevated, assess neurologic status every 15 minutes for 4 hours. If the patient is stable, assess every hour for 2 hours, then every 4 hours or according to pre-test schedule.
- Watch for complications of lumbar puncture, such as reaction to the anesthetic, meningitis, bleeding into the spinal canal, and cerebellar tonsillar herniation and medullary compression. Signs of meningitis include fever, neck rigidity, and irritability; of herniation, decreased level of consciousness, changes in pupil size and equality, and altered vital signs, including increased blood pressure, decreased pulse rate, and irregular respirations.

Peritoneal Fluid Analysis

ELEMENT	NORMAL VALUE OR FINDING
Gross appearance	Sterile, odorless, clear-to-pale yellow color; scant amount (< 50 ml)
RBCs	None
WBCs	< 300/µl
Protein	0.3 to 4.1 g/dl (albumin, 50% to 70%; globulin, 30% to 45%; fibrinogen, 0.3% to 4.5%)
Glucose	70 to 100 mg/dl
Amylase	138 to 404 amylase units/liter
Ammonia	< 50 µg/dl
Alkaline phosphatase	Male: > age 18, 90 to 239 units/liter Female: < age 45, 76 to 196 units/liter > age 45, 87 to 250 units/liter
Cytology	No malignant cells present
Bacteria	None
Fungi	None

Evaluating Peritoneal Fluid

While assisting with a peritoneal lavage, you can learn something about the patient's condition just by observing the appearance of his peritoneal fluid. Normal peritoneal fluid is clear to pale yellow. Here's what your patient's abnormal peritoneal fluid may indicate.

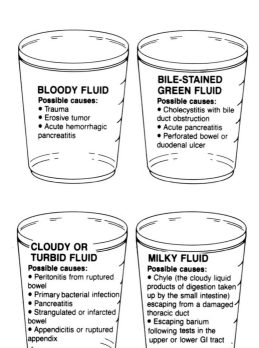

BLOODY FLUID
Possible causes:
- Trauma
- Erosive tumor
- Acute hemorrhagic pancreatitis

BILE-STAINED GREEN FLUID
Possible causes:
- Cholecystitis with bile duct obstruction
- Acute pancreatitis
- Perforated bowel or duodenal ulcer

CLOUDY OR TURBID FLUID
Possible causes:
- Peritonitis from ruptured bowel
- Primary bacterial infection
- Pancreatitis
- Strangulated or infarcted bowel
- Appendicitis or ruptured appendix

MILKY FLUID
Possible causes:
- Chyle (the cloudy liquid products of digestion taken up by the small intestine) escaping from a damaged thoracic duct
- Escaping barium following tests in the upper or lower GI tract

Synovial Fluid Analysis in Arthritis

Synovial fluid is normally a viscid, colorless-to-pale yellow liquid found in small amounts in the diarthrodial (synovial) joints, bursae, and tendon sheaths. It's thought to be produced by the dialysis of plasma across the synovial membrane and by the secretion of hyaluronic acid, a mucopolysaccharide. Although its functions aren't clearly understood, synovial fluid probably lubricates the joint space and nourishes the articular cartilage.

DISEASE	COLOR	CLARITY	VISCOSITY
Group I: noninflammatory			
Traumatic arthritis	Straw to bloody to yellow	Transparent to cloudy	Variable
Osteoarthritis	Yellow	Transparent	Variable
Group II: inflammatory			
Systemic lupus erythematosus	Straw	Clear to slightly cloudy	Variable
Rheumatic fever	Yellow	Slightly cloudy	Variable
Gout	Yellow to milky	Cloudy	Low
Rheumatoid arthritis	Yellow to green	Cloudy	Low
Group III: septic			
Septic arthritis	Gray or bloody	Turbid, purulent	Low

SYNOVIAL FLUID

In synovial fluid aspiration, or arthrocentesis, a sterile needle is inserted into a joint space—most commonly the knee—under strict aseptic conditions, to obtain a fluid specimen for analysis. This procedure is indicated in patients with undiagnosed articular disease and symptomatic joint effusion—the excessive accumulation of synovial fluid.

MUCIN CLOT	WBC COUNT/ % NEUTRO- PHILS	CARTILAGE DEBRIS/ CRYSTALS	RA CELLS/ BACTERIA
Good to fair	1,000/25%	None/ none	None/ none
Good to fair	700/15%	Usually present/ none	None/ none
Good to fair	2,000/30%	None/ none	LE cells/ none
Good to fair	14,000/50%	None/ none	LE cells may be present/ none
Fair to poor	20,000/70%	None/ urate	None/ none
Fair to poor	20,000/70%	None/ occasionally, cholesterol	Usually present/ none
Poor	90,000/90%	None/ none	None/ usually present

Adapted with permission from R. Jessar, "Synovianalysis in Arthritis," in Donald J. McCarty, *Arthritis and Allied Conditions: A Textbook of Rheumatology* (8th ed.; Philadelphia: Lea & Febiger, 1972).

Normal Findings in Synovial Fluid

ANALYSIS	RESULTS
Gross	
Color	Colorless to pale yellow
Clarity	Clear
Quantity (in knee)	0.3 to 3.5 ml
Viscosity	5.7 to 1,160
pH	7.2 to 7.4
Mucin clot	Good
Microscopic	
WBC count	0 to 200/µl
WBC differential	
• Lymphocytes	0 to 78/µl
• Monocytes	0 to 71/µl
• Clasmatocytes	0 to 26/µl
• Polymorphonuclears	0 to 25/µl
• Other phagocytes	0 to 21/µl
• Synovial lining cells	0 to 12/µl
Microbiologic	
Formed elements	Absence of cartilage debris and crystals
Bacteria	None
Serologic	
Complement	
• for 10 mg protein/dl	3.7 to 33.7 units/ml
• for 20 mg protein/dl	7.7 to 37.7 units/ml
Chemical	
Total protein	10.7 to 21.3 mg/dl
Fibrinogen	None
Glucose	70 to 100 mg/dl
Uric acid	2 to 8 mg/dl (men)
	2 to 6 mg/dl (women)
Hyaluronate	0.3 to 0.4 g/dl
$PaCO_2$	40 to 60 mm Hg
PaO_2	40 to 80 mm Hg

How Urine Is Formed

The body excretes urine only after several processes take place within the convoluted tubules. These processes include selective tubular reabsorption of water and solutes as well as selective tubular excretion of solutes.

This functional drawing and the charts on pages 42 and 43 should help explain where most products are reabsorbed and/or excreted during the process of urine formation. Note that the collecting tubules may reabsorb *or* excrete sodium, potassium ions, hydrogen ions, ammonium ions, and urea, depending on the patient's physiologic needs.

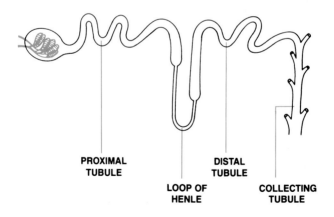

Continued

How Urine Is Formed
Continued

PROXIMAL TUBULE	LOOP OF HENLE	DISTAL TUBULE	COLLECTING TUBULE
Products reabsorbed			
-- Chloride ions	-- Sodium ions	-- Chloride ions	-- Water*
— Glucose	— Chloride ions	— Sodium ions +	— Sodium ions
— Potassium ions		-- Water*	— Potassium ions
— Amino acids		— Bicarbonate ions	— Hydrogen ions
— Bicarbonate ions			-- Ammonium ions
— Phosphate ions			-- Urea
-- Urea			
— Sodium ions			
-- Water			
— Uric acid			
— Magnesium ions			
— Calcium ions			

KEY   -- passive transport
       — active transport

* Antidiuretic hormone required for absorption to occur
+ Stimulated by aldosterone, which is controlled by the renin-angiotensin system

Continued

Nursing Tip

Hook the Bag: Have trouble ambulating a Foley catheter patient by yourself? Ask a hospital maintenance man to put a hook on the lower part of each movable I.V. pole. The patient can hold on to the I.V. pole for stability— and you can ambulate him by yourself.

Make sure the tubing doesn't hang low between the patient and pole, for the tubing could get caught under the pole's wheels.

—Linda Barker, RN

How Urine Is Formed
Continued

PROXIMAL TUBULE	LOOP OF HENLE	DISTAL TUBULE	COLLECTING TUBULE
Products excreted			
— Hydrogen ions	-- Sodium chloride	— Water	— Urea
— Foreign substances		— Potassium ions +	— Sodium ions
— Creatinine		-- Urea	— Potassium ions
		— Hydrogen ions	— Hydrogen ions
		-- Ammonium ions	-- Ammonium ions
		— Uric acid	
Character of filtrate			
Isotonic	Hypotonic	Isotonic or hypotonic	Concentrated

KEY   --   passive transport
       —   active transport

* Antidiuretic hormone required for absorption to occur
+ Stimulated by aldosterone, which is controlled by the renin-angiotensin system

Urinary Output: Rule of Thumb

Daily intake made up of visible fluids (1,200 ml), plus three meals (1,000 ml), plus water of oxidation (300 ml) should exceed the daily urine output (1,500 ml) by at least 500 ml. Minimum urine output in an adult should be at least 600 ml.

The Particulars of Concentration

An isotonic fluid has a concentration of dissolved particles, or tonicity, equal to that of the intracellular fluid. When isotonic fluids such as 5% dextrose in water or 0.9% sodium chloride enter the circulation, they cause no net movement of water across the semipermeable cell membrane. And because the osmotic pressure is the same inside and outside the cells, they do not swell or shrink.

A hypertonic fluid has a concentration of dissolved particles greater than that of intracellular fluid. When a hypertonic solution such as 3% sodium chloride or 50% dextrose is rapidly infused into the body, water will rush out of the cells to the area of greater concentration of particles and the cells will shrivel. Dehydration can also make extracellular fluid hypertonic.

A hypotonic fluid has a concentration of dissolved particles less than that of intracellular fluid. When a hypotonic solution such as distilled water surrounds a cell, water will diffuse into the intracellular fluid, causing the cell to swell. Inappropriate use of intravenous fluids or severe electrolyte loss will make body fluids hypotonic.

Evaluating Urine Color

APPEARANCE	POSSIBLE CAUSES
Colorless or straw-colored (diluted urine)	• Excessive fluid intake, chronic renal disease, diabetes insipidus, nervous conditions
Dark yellow or amber (concentrated urine)	• Low fluid intake, acute febrile disease, vomiting, diarrhea
Cloudy	• Infection, purulence, blood, epithelial cells, fat, colloidal particles, urates, vegetarian diet, parasitic disease
Yellow to amber, with pink sediment	• Hyperuricemia, gout
Orange-red to orange-brown	• Urobilinuria, such drugs as phenazopyridine (Pyridium), obstructive jaundice (tea-colored)
Red or red-brown	• Porphyria, hemoglobin, erythrocytes, hemorrhage, such drugs as pyrvinium pamoate (Povan)
Green-brown	• Bile duct obstruction, phenol poisoning
Dark brown or black	• Acute glomerulonephritis, chorea, typhus, methylene blue medication
Smoky	• Prostatic fluid, fat droplets, blood, chyle, spermatozoa

Analyzing a Urine Specimen

CHARACTERISTIC	NORMAL FINDING
Turbidity (lucidity)	Clear
Specific gravity (concentration as compared with H_2O)	1.010 to 1.025 (varies with hydration level)
Acidity (pH)	pH 4.8 to 7.5
Protein	Little to no protein
Glucose	None
Red blood cells	None to three per high-power field
White blood cells	None to four per high-power field
Casts	None

ABNORMAL FINDINGS	POSSIBLE CAUSES
• Cloudy	• Infection • Urinary tract disease • Nonpathologic causes such as presence of amorphous phosphate
• Dilute (1.001 to 1.010) or concentrated (1.025 to 1.030) urine	• Low specific gravity may be caused by renal disorders or diabetes insipidus. • Nonpathologic causes for high specific gravity include administration of albumin, contrast media, or dextran.
• Greater than pH 7.5	• Renal acidosis • Calculi
• Proteinuria	• Renal or proximal tubule disorder • Toxemia • Severe heart failure
• Glycosuria	• Gestational diabetes • Diabetes mellitus • Nonpathologic causes such as dextrose therapy
• More than three per high-power field	• Renal malfunction or trauma to the urinary tract • Urinary tract tumor or infection
• More than four per high-power field	• Bacterial infection
• Clear or yellowish, cylindrically shaped cellular debris, visible under a microscope	• Renal damage

Drugs That Affect Routine Urinalysis Results

DRUGS THAT CHANGE URINE COLOR:

Alcohol (light, due to diuresis)
Chlorpromazine hydrochloride (dark)
Chlorzoxazone (orange to purple-red)
Deferoxamine mesylate (red)
Fluorescein sodium I.V. (yellow-orange)
Furazolidone (brown)
Levodopa (dark)
Methylene blue (blue-green)
Metronidazole (dark)
Nitrofurantoin (brown)
Oral anticoagulants, indanedione derivatives (orange)
Phenazopyridine (orange-red, orange-brown, or red)
Phenolphthalein (red to purple-red)
Phenolsulfonphthalein (pink or red)
Quinacrine (deep yellow)
Riboflavin (yellow)
Rifampin (red-orange)
Sulfasalazine (orange-yellow)
Sulfobromophthalein (red)

DRUGS THAT CAUSE URINE ODOR:

Antibiotics
Paraldehyde
Vitamins

DRUGS THAT INCREASE SPECIFIC GRAVITY:

Albumin
Dextran
Glucose
Radiopaque contrast media

DRUGS THAT DECREASE pH:

Ammonium chloride
Diazoxide
Methenamine
Metolazone

DRUGS THAT INCREASE pH:

Acetazolamide
Amphotericin B
Mafenide
Sodium bicarbonate
Potassium citrate

DRUGS THAT SHOW A FALSE-POSITIVE RESULT FOR PROTEINURIA:

Acetazolamide (Combistix)
Aminosalicylic acid (sulfosalicylic acid or Exton's method)
Cephaloridine and cephalothin (sulfosalicylic acid method)
Nafcillin (sulfosalicylic acid method)
Sodium bicarbonate (all methods)
Tolbutamide (sulfosalicylic acid method)
Tolmetin (sulfosalicylic acid method)   *Continued*

Drugs That Affect Routine Urinalysis Results
Continued

DRUGS THAT CAUSE TRUE PROTEINURIA:

Amphotericin B
Bacitracin
Gentamicin
Gold preparations
Kanamycin
Neomycin
Phenylbutazone
Polymyxin B
Streptomycin
Trimethadione

DRUGS THAT CAN CAUSE EITHER TRUE PROTEINURIA OR FALSE-POSITIVE RESULTS:

Penicillin in large doses (except with Ames Reagent Strips); however, some penicillins cause true proteinuria
Sulfonamides (sulfosalicylic acid method)

DRUGS THAT CAUSE FALSE-POSITIVE GLYCOSURIA:

Aminosalicylic acid (Benedict's test)
Ascorbic acid (Clinistix, Diastix, or Tes-Tape)
Ascorbic acid in large doses (Clinitest tablets)
Cephalosporins (Clinitest tablets)
Chloral hydrate (Benedict's test)
Chloramphenicol (Benedict's test or Clinitest tablets)
Isoniazid (Benedict's test)
Levodopa (Clinistix, Diastix, or Tes-Tape)
Levodopa in large doses (Clinitest tablets)
Methyldopa (Tes-Tape)
Nalidixic acid (Benedict's test or Clinitest tablets)
Nitrofurantoin (Benedict's test)
Penicillin G in large doses (Benedict's test)
Phenazopyridine (Clinistix, Diastix, or Tes-Tape)
Probenecid (Benedict's test or Clinitest tablets)
Salicylates in large doses (Clinitest tablets, Clinistix, Diastix, or Tes-Tape)
Streptomycin (Benedict's test)
Tetracycline (Clinistix, Diastix, Tes-Tape)
Tetracyclines (Benedict's test or Clinitest tablets)

DRUGS THAT CAUSE TRUE GLYCOSURIA:

Ammonium chloride
Asparaginase
Carbamazepine
Corticosteroids
Dextrothyroxine
Lithium carbonate
Nicotinic acid (large doses)
Phenothiazines (long-term)
Thiazide diuretics

Continued

Drugs That Affect Routine Urinalysis Results
Continued

DRUGS THAT INCREASE WBC:

Allopurinol
Ampicillin
Aspirin toxicity
Kanamycin
Methicillin

DRUGS THAT CAUSE HEMATURIA:

Amphotericin B
Coumarin derivatives
Methenamine in large doses
Methicillin
Para-aminosalicylic acid
Phenylbutazone
Sulfonamides

DRUGS THAT CAUSE FALSE-POSITIVE RESULTS FOR KETONURIA:

Levodopa (Ketostix or Labstix)
Phenazopyridine (Ketostix or Gerhardt's reagent strip shows atypical color)
Phenolsulfonphthalein (Rothera's test)
Phenothiazines (Gerhardt's reagent strip shows atypical color)
Salicylates (testing with Gerhardt's reagent strip shows reddish color)
Sulfobromophthalein (Bili-Labstix)

DRUGS THAT CAUSE TRUE KETONURIA:

Ether—anesthesia
Isoniazid—intoxication
Isopropyl alcohol—intoxication
Insulin—excessive doses

DRUGS THAT CAUSE CASTS:

Amphotericin B
Aspirin toxicity
Bacitracin
Ethacrynic acid
Furosemide
Gentamicin
Griseofulvin
Isoniazid
Kanamycin
Neomycin
Penicillin
Radiographic agents
Streptomycin
Sulfonamides

DRUGS THAT CAUSE CRYSTALS (IF URINE IS ACIDIC):

Acetazolamide
Aminosalicylic acid
Ascorbic acid
Nitrofurantoin
Theophylline
Thiazide diuretics

Analyzing Urine Specimens

SUBSTANCE/NORMAL LEVELS IN URINE	IMPLICATIONS OF ABNORMAL LEVELS
Proteins **Protein** Less than 150 mg/24 hr	↑ Increased glomerular permeability, urinary tract disorders, renal disease, leukemia and severe anemias, toxemia, abdominal tumors, intestinal obstruction, cardiac disease, poisoning
Bence Jones protein Negative	↑ Multiple myeloma
Protein metabolites **Amino acids** 50 to 200 mg/24 hr	↑ Defective tubular reabsorption, resulting in increased renal excretion; congenital enzyme deficiencies, leading to metabolic disorders, such as phenylketonuria; maple sugar disease; cystinuria; homocystinuria, argininosuccinicaciduria, histidinemia, hyperprolinemia Type A, citrullinuria
Creatinine Clearance in men (age 20): 90 ml/min/1.73m^2 Clearance in women (age 20): 84 ml/min/1.73m^2 (Concentrations decrease 6 ml/min/decade in older patients)	↓ Reduced renal blood flow, acute tubular necrosis, acute or chronic glomerulonephritis, advanced bilateral chronic pyelonephritis, advanced bilateral renal lesions, nephrosclerosis, severe dehydration, congestive heart failure
Urea Maximal clearance: 64 to 99 ml/min	↓ Reduced renal blood flow, glomerulonephritis, bilateral chronic pyelonephritis, tubular necrosis, nephrosclerosis, bilateral renal lesions, bilateral ureteral obstruction, CHF, dehydration

Continued

Analyzing Urine Specimens
Continued

SUBSTANCE/NORMAL LEVELS IN URINE	IMPLICATIONS OF ABNORMAL LEVELS
Protein metabolites **Uric acid** 250 to 750 mg/24 hr (varies with purine intake)	↑ Chronic myeloid leukemia, polycythemia vera, multiple myeloma, pernicious anemia, lymphosarcoma and lymphatic leukemia during radiotherapy, defective tubular reabsorption, Wilson's disease
	↓ Chronic glomerulonephritis, diabetic glomerulosclerosis, collagen disorders
Pigments **Hemoglobin** Negative	↑ Severe intravascular hemolysis due to transfusion reaction, burns, or crushing injury; acquired or congenital hemolytic anemias. Less commonly: cystitis, ureteral calculi, urethritis. With hematuria: acute glomerulonephritis or pyelonephritis, renal tumor, tuberculosis
Myoglobin Negative	↑ Severe muscular damage from burns or injury, acute or chronic muscular disease, alcoholic polymyopathy, familial myoglobinuria, extensive myocardial infarction
Delta-aminolevulinic acid (ALA) 1.5 to 7.5 mg/dl/24 hr	↑ Lead poisoning, acute porphyria, hepatic carcinoma, hepatitis
Porphobilinogen (PBG) up to 1.5 mg/24 hr	↑ Acute intermittent porphyria, carcinomatosis, hepatitis

Continued

Analyzing Urine Specimens
Continued

SUBSTANCE/NORMAL LEVELS IN URINE	IMPLICATIONS OF ABNORMAL LEVELS
Pigments **Uroporphyrin** *Men:* 0 to 42 mcg/24 hr *Women:* 1 to 22 mcg/24 hr	↑ Cirrhosis of the liver, hemochromatosis, lead poisoning, acute porphyria
Coproporphyrin *Men:* 0 to 96 mcg/24 hr *Women:* 1 to 57 mcg/24 hr	↑ Alcoholic cirrhosis, chemical toxicity, increased erythropoietic activity accompanying anemia, infectious hepatitis, lead poisoning, malignant conditions, myocardial infarction, obstructive jaundice, acute porphyrias
Bilirubin Negative	↑ Hepatic disease due to infectious or toxic agents, obstructive biliary disease
Urobilinogen *Men:* 0.3 to 2.1 Ehrlich units/2 hr *Women:* 0.1 to 1.1 Ehrlich units/2 hr	↑ Hemolytic jaundice, hepatitis, cirrhosis, drug treatment that alkalizes urine ↓ Congenital enzymatic jaundice, drug treatment that acidifies urine Absence may indicate complete obstructive jaundice; broad-spectrum antibiotic treatment
Melanin Negative	↑ Malignant melanoma

Analyses of abnormal urine levels of proteins, protein metabolites, and pigments help identify and determine the treatment of renal, extrarenal, and systemic diseases.

Analyzing a 24-Hour Urine Specimen

When you study this chart, note the creatinine clearance value. This value isn't calculated over a 24-hour period, as the other values are. Instead, using the patient's body surface and age, it's an estimate of the amount of creatinine that the kidney can filter in 1 minute.

TEST/NORMAL VALUE	IMPLICATION OF ABNORMALITY
Aldosterone *Male and female:* 2 to 16 mcg/24 hr	Increase indicates either primary or secondary aldosteronism
17-hydroxycorticosteroids *Male and female:* 3.8 mg/24 hr	Decrease indicates adrenocortical malfunction.
17-ketosteroids *Male:* 6 to 21 mg/24 hr *Female:* 4 to 17 mg/24 hr	Increase indicates adrenal hyperplasia, carcinoma, adenoma, or adrenogenital syndrome.

Catecholamines (norepinephrine, epinephrine, dopamine)
Male and female:
  Norepinephrine: 15 to 80 mcg/24 hr
  Epinephrine: 0 to 20 mcg/24 hr
  Dopamine: 65 to 400 mcg/24 hr
	Increase in any may indicate pheochromocytoma.

Calcium, phosphorus, and creatinine clearance
Calcium
  Male: < 275 mg/24 hr
  Female: < 250 mg/24 hr
Phosphorus
  Male and female: < 1,100 mg/24 hr
Creatinine
  Male: 90 ml/min/1.73 m^2 at age 20
  Female: 84 ml/min/1.73 m^2 at age 20
(Decrease by 6 ml/min/decade)
	Increase in any indicates hyperparathyroidism, which can cause calculi.

GENERAL INFORMATION

Assessing Tissue Perfusion: Tests That Help

The following tests will help you assess tissue perfusion, electrolyte levels and acid-base status, the need for sodium bicarbonate, and the patient's response to treatment.

VENOUS BLOOD

hematocrit
protein
electrolytes (Na, K, Cl, HCO_3^-)
CO_2
pH
PO_2
urea (BUN)

URINARY OUTPUT

Na, K, and Cl concentration
specific gravity
protein
sugar and acetone
pH
color
blood

Principal Extracellular Electrolytes

The principal extracellular electrolytes are sodium, calcium, and bicarbonate. The principal intracellular electrolytes are potassium, magnesium, and phosphate. Sodium is the dominant extracellular cation (+); potassium, the dominant *intracellular* cation. Chloride is the dominant extracellular anion (−); phosphate, the dominant *intracellular* anion. Each fluid compartment has its own composition of electrolytes, and these must be in the right compartments in the right amounts. The unit of measure used for these electrolytes is the milliequivalent (mEq).

Serum Electrolyte Functions

CATIONS

Sodium (Na$^+$)
- Maintains osmotic pressure of extracellular fluid
- Helps regulate neuromuscular activity
- Influences acid-base balance, and chloride and potassium levels
- Helps regulate water excretion

Potassium (K$^+$)
- Maintains intracellular osmotic equilibrium
- Helps regulate neuromuscular and enzymatic activity, and acid-base balance
- Influences kidney function

Calcium (Ca^{++})
- Helps regulate and promote neuromuscular activity, skeletal development, and blood coagulation

Magnesium (Mg^{++})
- Helps regulate intracellular activity, and sodium, potassium, calcium, and phosphorous levels

ANIONS

Chloride (Cl$^-$)
- Influences acid-base balance
- Helps maintain blood osmotic pressure and arterial pressure

Bicarbonate (HCO$_3^-$)
- Acts with carbonic acid in buffer system that regulates blood pH

Phosphate (HPO$_4^{--}$)
- Helps regulate calcium levels, energy metabolism, and acid-base balance

Electrolyte Sources

MICRONUTRIENT/ FOOD SOURCES	DISORDERS
Sodium Table salt, beef, pork, sardines, cheese, milk, eggs	**Toxicity:** hypernatremia **Deficiency:** hyponatremia
Chloride Table salt, seafood, milk, meat, eggs	**Toxicity:** hyperchloremia **Deficiency:** hypochloremia
Potassium Milk, dates, meat, fish, bananas	**Toxicity:** hyperkalemia **Deficiency:** hypokalemia
Calcium Milk, milk products, meat, fish, eggs, cereals, beans, fruit, vegetables	**Toxicity:** hypercalcemia **Deficiency:** hypocalcemia
Phosphorus Milk, cheese, meat, poultry, fish, whole-grain cereals, nuts, legumes	**Toxicity:** hyperphosphatemia **Deficiency:** hypophosphatemia
Magnesium Seafood, soybeans, nuts, cocoa, whole-grain cereals, peas, dried beans, meat, milk	**Toxicity:** hypermagnesemia **Deficiency:** hypomagnesemia
Iron Liver, meat, egg yolks, beans, clams, peaches, whole or enriched grains, legumes	**Toxicity:** hemochromatosis **Deficiency:** anemia

Drug Interactions with Electrolytes

The chart below lists several drugs and drug types that may affect your patient's potassium, magnesium, or calcium levels. Although these electrolyte level changes may be therapeutic in some instances, they can place the patient at risk of a potentially dangerous electrolyte imbalance. If your patient's receiving any of these drugs, closely monitor his serum electrolyte levels and report imbalances.

DRUG	POSSIBLE EFFECTS
Acetazolamide	Hypocalcemia
Alcohol	Hypomagnesemia
Aminoglycosides	Hypomagnesemia, hypokalemia
Aminosalicylic acid	Hypokalemia
Amphotericin B	Hypokalemia, hypomagnesemia
Bumetanide	Hypomagnesemia, hypocalcemia, hypokalemia
Capreomycin	Hypomagnesemia, hypocalcemia
Cisplatin	Hypomagnesemia
Corticosteroids	Hypokalemia, hypocalcemia
Cycloserine	Hypomagnesemia, hypocalcemia
Dactinomycin	Hypocalcemia
Ethacrynic acid	Hypomagnesemia, hypocalcemia, hypokalemia
Furosemide	Hypomagnesemia, hypocalcemia, hypokalemia
Glutethimide	Hypocalcemia

Continued

Drug Interactions with Electrolytes
Continued

DRUG	POSSIBLE EFFECTS
Levodopa	Hypokalemia
Lithium	Hypermagnesemia, hypocalcemia
Mercurials	Hypomagnesemia, hypocalcemia
Oral contraceptives	Hypomagnesemia
Penicillins	Hypokalemia
Probenecid	Hypokalemia, hypomagnesemia, hypocalcemia
Salicylates	Hypokalemia
Spironolactone	Hypomagnesemia, hypocalcemia, hyperkalemia
Tetracyclines	Hypomagnesemia, hypocalcemia
Thiazides	Hypomagnesemia, hypercalcemia, hypokalemia
Triamterene	Hypocalcemia, hyperkalemia

Normal Urine and Serum Sodium Values

Normal urine sodium values
Normal urine sodium excretion is 30 to 280 mEq/24 hours; normal urine chloride excretion, 110 to 250 mEq/24 hours; and normal urine sodium chloride excretion, 5 to 20 g/24 hours.

Normal serum sodium values
Normally, serum sodium levels range from 135 to 145 mEq/liter.

Why Is Sodium Important?

Sodium is the most abundant cation in extracellular fluid. At its normal concentration, 135 to 145 mEq/liter, it supplies almost 90% of the total cations. Sodium affects many vital functions and is mainly responsible for the osmotic pressure of the extracellular fluid. This is partly because of its prevalence in the body and partly because it does not easily cross the cell membrane.

Discussing body sodium without also mentioning water is almost impossible. The sodium concentration of the extracellular fluid profoundly influences the kidneys' regulation of the body's water and electrolyte status. For example, when sodium concentration falls, the kidneys promote water excretion under the influence of aldosterone and other stimuli; when sodium concentration rises, the release of ADH causes the kidneys to retain more water and subsequently dilute the sodium to normal levels.

Sodium promotes the irritability of nerve and muscle tissue and the conduction of nerve impulses and influences the body's vital acid-base balance.

Generally, the body's regulatory mechanisms hold the sodium-water relationship within normal limits. But illness can throw this relationship severely out of balance.

Feedback Mechanism Between Sodium and Aldosterone

Decreased extracellular concentration of sodium, increased extracellular concentration of potassium, or decreased cardiac output stimulates adrenocorticotropic hormone (ACTH) release from the anterior pituitary, which, in turn, increases aldosterone secretion from the adrenal cortex. Aldosterone promotes renal retention of sodium and water and excretion of potassium to increase extracellular sodium and water and decrease extracellular potassium.

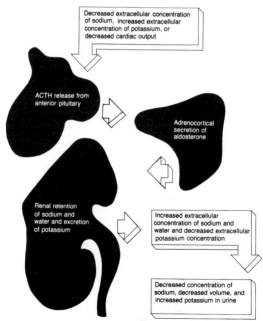

Hyponatremia

HYPONATREMIA (serum sodium below 135 mEq/liter)

Causes
Hyponatremia associated with excess fluid
- SIADH (syndrome of inappropriate secretion of ADH)
- Excessive intake of hypotonic fluids

Hyponatremia associated with dehydration
- Administration of salt-removing diuretics
- Salt-wasting nephritis
- Excessive gastrointestinal fluid losses from nasogastric suctioning, vomiting, or diarrhea
- Severe diaphoresis
- Potassium depletion
- Trauma, such as burns or wound drainage
- Aldosterone deficiency
- Severe malnutrition

Signs and symptoms
Hyponatremia associated with excess fluid
- Headache, anxiety, lassitude, apathy, confusion
- Anorexia, nausea, vomiting, diarrhea, cramping
- Hyperreflexia, muscle spasms, muscle weakness

Hyponatremia associated with dehydration
- Irritability, tremors, seizures, coma
- Dry mucous membranes
- Low-grade fever
- Hypotension, tachycardia
- Decreased urine output ranging from oliguria to anuria

Nursing considerations
- Correct sodium and water imbalance through diet and administration of I.V. solutions, as ordered.
- During administration of saline solution, observe patient closely for signs of hypervolemia (such as dyspnea, crackles, and engorged neck or hand veins).
- During treatments, monitor neurologic and gastrointestinal symptoms to detect improvement or deterioration.
- Monitor sodium and potassium levels closely.

Continued

Hypernatremia
Continued

HYPERNATREMIA (serum sodium above 145 mEq/liter)

Causes
Hypernatremia associated with excess fluid
- Excessive administration of large amounts of sodium chloride solution I.V.
- Excessive aldosterone secretion

Hypernatremia associated with dehydration
- Dehydration *without sodium loss* from decreased water intake, severe vomiting, or diarrhea
- Excessive administration of osmotic diuretics
- Hypercalcemia with polyuria and dehydration
- Neurohypophyseal dysfunction (as in diabetes insipidus)
- Renal tubular disease
- Dysfunctional thirst mechanism

Signs and symptoms
Hypernatremia associated with excess fluid
- Weight gain
- Pitting edema in extremities
- Hypertension
- Shortness of breath (only in severe imbalance)
- Agitation, restlessness, convulsions (only in severe imbalance)

Hypernatremia associated with dehydration
- Lethargy, irritability, tremors, seizures, coma
- Dry mucous membranes; rough, dry tongue; flushed skin
- Low-grade fever
- Oliguria
- Intense thirst

Nursing considerations
- Monitor hourly urine output.
- Replace water volume, as ordered, administering fluids with caution.
- Check serum sodium levels every 6 hours.
- Monitor vital signs closely. Watch for increasing pulse rate and hypertension.
- Observe patient for signs of hypervolemia.

The Sodium Pump

The processes of depolarization and repolarization cause sodium to diffuse into the cells and potassium to diffuse out of the cells, disrupting the ionic balance. Although sodium and potassium don't easily penetrate the cell membrane, in combination with a carrier they can become soluble in the lipoprotein membrane and can be transported across to their original compartments.

A single carrier is thought to transport both ions. This carrier is called Y when it has an affinity for sodium and X after it undergoes a chemical transformation which changes its affinity, allowing it to combine with potassium. Adenosine triphosphate (ATP) and ATPase provide the energy needed for this transformation as well as for splitting the cations from the carrier at the appropriate moment. The entire cycle goes on indefinitely and is referred to as the sodium pump. It helps transmit neuromuscular impulses, stimulate glandular secretions, and prevent cellular swelling.

Not Worth the Salt

Instruct your edematous patient to avoid all foods high in sodium, which include the following:
- salted snack foods, such as potato chips and peanuts
- canned soups and vegetables
- dried fruits
- delicatessen foods, especially lox
- prepared pre-portioned foods, such as TV dinners
- preserved meats (such as hot dogs) and luncheon meats
- cheeses of all kinds (including cottage)
- anything preserved in brine, such as olives, pickles, and sauerkraut

What's Cooking?

Here are some helpful food suggestions for the patient with fluid problems controlled by a diuretic. These foods are high in potassium but low in sodium.

Fruits and their juices
Apples
Apricots
Bananas
Cherries
Currants
Dates
Grapefruit
Mangoes
Nectarines
Oranges
Pears
Peaches
Pineapples
Plums
Prunes
Raisins
Raspberries
Strawberries
Watermelon

Vegetables (fresh or frozen)
Asparagus
Beans
Brussels sprouts
Cabbage
Cauliflower
Corn
Lima beans
Peas
Peppers
Potatoes
Radishes
Squash

Cereals and starches
Oatmeal
Wheat germ

Some Sodium Supplements

Hyponatremia simply means that too little sodium compared to water is present in the blood. Several things can cause this deficiency, including water intoxification and SIADH (syndrome of inappropriate secretion of ADH). Depending upon the cause, sodium supplementation may be ordered. Here's a list of products the doctor may choose.

PRODUCT	AMOUNT OF SODIUM SUPPLIED	NURSING TIPS
Tablets		
Sodium chloride	300 mg, 500 mg, 600 mg, 650 mg, 1 g, 2.25 g, 2.5 g Enteric-coated: 1 g	• Enteric-coated tablets must not be crushed, chewed or dissolved
Sodium chloride with dextrose	455 mg sodium chloride and 195 mg dextrose (also with Vitamin B_1)	
Thermotabs	450 mg sodium chloride, 30 mg potassium chloride, 18 mg calcium carbonate, 200 mg dextrose	
Oral solution		
Moyer's solution	Sodium chloride 0.3% and sodium bicarbonate 0.15%	• Especially useful to replace fluids after burns and injuries

Continued

Some Sodium Supplements
Continued

PRODUCT	AMOUNT OF SODIUM SUPPLIED	NURSING TIPS
Parenteral		
Sodium chloride	0.45% solution (0.45 g/100 ml) 0.9% solution (0.9 g/100 ml) 3% solution (3 g/100 ml) 5% solution (5 g/100 ml)	• Watch for signs of electrolyte imbalance: water retention, edema, and aggravation of hypokalemia, or acidosis. Hypertonic solutions 3% and 5% can cause increased venous pressure. Administer cautiously in small quantities by slow I.V. • In infants, don't exceed 8 mEq/kg/day to avoid hypernatremia, decreased cerebrospinal fluid pressure, and intracranial hemorrhage.
Ringer's lactate solution	130 mEq/liter of sodium chloride and 27 mEq/liter of sodium lactate (plus potassium and calcium)	• Give sodium solutions cautiously to patients with CHF, kidney dysfunction, or circulatory insufficiency.
Sodium bicarbonate	4.2% (500 mEq/liter) in 10 ml 5% (595 mEq/liter) in 500 ml 7.5% (892 mEq/liter) in 50 ml 8.4% (1,000 mEq/liter) in 50 ml and 10 ml	• Usually given rapidly during cardiac arrest. In less urgent metabolic acidosis, infuse hypertonic $NaHCO_3$ at 2 to 5 mEq/kg over 4 to 8 hours to avoid overcorrection in alkalosis.

Sodium Content of Commonly Used OTC Products

Antacids/Antiflatulents
Riopan-Magaldrate	<0.1 mg/tsp
Maalox	1.35 mg/tsp
Maalox plus	1.3 mg/tsp
Maalox TC	0.8 mg/tsp
Gaviscon	12.9 mg/tsp

Laxatives/Stool Softeners
Surfak	none
Metamucil	negligible
Metamucil (packets)	high
Konsyl	none
Feen-a-mint	none
Dulcolax Suppositories	none
Fletcher's Castoria	none
Haley's M-O	high
Milk of Magnesia	high
Fleet's	high
Senokot	high
Colace	negligible

Cough Preparations
Robitussin	none
Robitussin A-C and DM	none
Silence is Golden	none
Romilar	none
Congespirin	none
Coricidin and Coricidin D	none
Vicks' Formula 44	none

Weight Control
Sweeta	none

Potassium: Normal Serum Values

Normally, serum potassium levels range from 3.5 to 5.5 mEq/liter.

Why Is Potassium Important?

Because it's the dominant intracellular electrolyte and thus controls cellular osmotic pressure. Further, potassium activates several enzymatic reactions; helps regulate acid-base balance; influences kidney function and structure (nephropathy can follow prolonged potassium deficit); and maintains neuromuscular excitability.

Potassium exists normally in the serum within a very narrow range—3.5 to 5.5 mEq/liter. Small deviations in either direction can have disastrous consequences, particularly for persons already ill. Even in those who are reasonably well, potassium imbalance can quickly become surprisingly severe.

Fortunately, potassium levels usually stay within normal limits despite a great fluctuation in fluid and electrolyte intake. Like other electrolytes, potassium constantly shifts among blood, cells, gastrointestinal fluids, and urine. The same shifting goes on in sweat and salivary glands as well. What influences this movement of potassium? Adrenal steroid hormones, changes in pH, and changes in blood glucose levels do. So do changes in blood sodium levels: There seems to be a reciprocal relation between sodium and potassium; *a large intake of sodium increases the loss of potassium, and vice versa.*

Hypokalemia

HYPOKALEMIA (serum potassium level below *3.5 mEq/liter*)

Causes
- Thiazide and osmotic diuretic therapy
- Prolonged potassium-free I.V. therapy (for a patient receiving nothing by mouth)
- Renal disease such as tubular acidosis and Fanconi's syndrome
- Excessive aldosterone secretion
- Acid-base imbalance
- Excessive gastrointestinal fluid losses from nasogastric suctioning, vomiting, diarrhea, or intestinal fistula
- Malnutrition or malabsorption syndrome
- Laxative abuse
- Trauma with associated loss of potassium in urine

Signs and symptoms
- Dysrhythmias, enhanced effectiveness of digitalis (to the point of toxicosis), presence of U wave and depressed ST segment on EKG waveform.
- Muscle weakness, fatigue, leg cramps
- Drowsiness, irritability, coma
- Anorexia, vomiting, paralytic ileus
- Polyuria

Nursing considerations
- Place patient on cardiac monitor, as ordered. Observe him for changes in heart rate, rhythm, and EKG pattern.
- Treat dysrhythmias by correcting potassium imbalance, as ordered; method depends on serum level and severity of symptoms. Doctor may order increased potassium intake in diet or oral potassium supplements (diluted to prevent irritation and to facilitate absorption); or, in emergency, slow administration of diluted potassium chloride I.V. (with patient on cardiac monitor). Monitor patient for signs and symptoms of sudden hyperkalemia onset.
- Monitor serum potassium level to determine effects of replacement therapy.
- Determine source of potassium loss; for example, wound or fistula drainage. Analysis of drainage sample electrolyte content will help indicate amount of daily loss.
- Observe patient for digitalis toxicosis.
- Observe patient for signs of alkalosis, such as irritability, confusion, diarrhea, and nausea.
- Never give potassium when patient's urine output is below 600 ml/day.

Hyperkalemia

HYPERKALEMIA (serum potassium level above 5.5 mEq/liter)

Causes
- Excessive administration of potassium chloride
- Renal disease
- Use of potassium-sparing diuretics by renal disease patients
- Destruction of cells by burns or trauma, with subsequent potassium release
- Aldosterone insufficiency
- Acidosis, for example, diabetic ketoacidosis

Signs and symptoms
- Cardiac symptoms, including bradycardia, lethal dysrhythmias, and cardiac arrest; EKG waveform shows tented T wave, prolonged PR interval, widened QRS complex, flat-to-absent P wave and asystole
- Apathy, confusion, tingling
- Hyperreflexia progressing to numbness, tingling, and flaccid weakness, paralysis
- Abdominal cramping, nausea, diarrhea
- Oliguria, anuria
- Metabolic acidosis

Nursing considerations
- Place patient on cardiac monitor, as ordered. Observe him for changes in heart rate, rhythm, and EKG pattern.
- Restrict ingestion of potassium in diet.
- In an emergency (when potassium excess causes EKG changes or the serum potassium level exceeds 6 mEq/liter), administer I.V. glucose, insulin, and sodium bicarbonate, as ordered. This treatment will help shift potassium into the cell. *Note:* Treatment's effects last about 4 hours.
- Check serum potassium levels to monitor shift of potassium in and out of cells. With I.V. treatment, potassium shifts rapidly into cells. When I.V. treatment is discontinued, potassium shifts back into blood.
- Follow I.V. treatment with one or both of the following, as ordered: sodium polystyrene sulfonate (Kayexalate*), an exchange resin, administered orally, through a nasogastric tube, or as an enema; or dialysis. If these treatments are prolonged, monitor patient for resulting hypokalemia.

*Available in both the United States and in Canada

Potassium Imbalances

The highs
When your patient's serum potassium level is above normal, expect the following EKG changes.

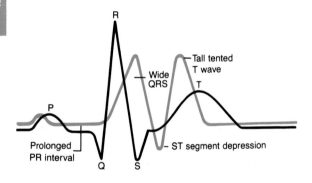

■ Normal   ■ Abnormal

Continued

Potassium Imbalances
Continued

The lows
When your patient's serum potassium level is below normal, expect the following EKG changes.

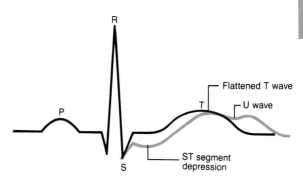

■ Normal   ■ Abnormal

Expect Hyperkalemia

You can expect hyperkalemia in the following situations that allow heavy excretion of sodium, release of intracellular potassium stores, and decreased excretion of potassium: Addison's disease, hemorrhagic shock and acute renal failure, massive crushing injury, and myocardial infarction.

Treatment for Hyperkalemia May Include:

• *Cation exchange resin* — (polystyrene sulfonate/Kayexalate) binds and eliminates potassium via the bowel. It can be given orally, rectally, or by nasogastric tube. A common mixture is 50 g of resin suspended in 50 ml of 70% Sorbitol and 100 ml of H_2O.

• *$NaHCO_3$* — intravenous infusion of 44 to 132 mEq $NaHCO_3$ added to 1 liter 5% dextrose in water. $NaHCO_3$ rapidly lowers plasma potassium levels by alkalinization of plasma, which causes a shift of K^+ back into the cell. The beneficial effect is also due partly to the dilution of the plasma K^+ by administration of a hypertonic sodium solution (i.e., 7.5% $NaHCO_3$).

• *Calcium salts* — by slow I.V. infusion of 4.5 to 13.5 mEq of 10% calcium gluconate under EKG monitoring. Calcium does not alter plasma potassium levels but acts on the neuromuscular membranes and antagonizes the cardiotoxicity of hyperkalemia (when EKG shows broad QRS complexes or absent P waves). The rate of calcium infusion should not exceed 1.5 mEq/min; the total daily infusion in adults should not exceed 70 mEq. The effects of calcium infusion are rapid but transient.

• *Glucose and insulin* — by I.V. infusion of 500 ml of 10% dextrose in water and 10 units of regular insulin over 30 minutes. This infusion is followed by a slower one of just dextrose in water.

• *Dialysis* — (hemo- or peritoneal dialysis) effectively removes potassium from the serum. However, these methods are relatively slow compared to other treatments. Dialysis is rarely required for treating electrolyte imbalance except in patients with renal failure.

Some Potassium Supplements

PRODUCT	AMOUNT OF POTASSIUM SUPPLIED◊	NURSING TIPS
Liquids		
Kaochlor 10%	10% (20 mEq/15 ml)	• Give these supplements with extreme caution. Never switch potassium products without a doctor's order. If your patient tolerates one product better than another, tell the doctor so he can change the brand and dosage.
Kaochlor S-F 10%	10% (20 mEq/15 ml)	
Kay Ciel*	10% (20 mEq/15 ml)	
Klor-10%	10% (20 mEq/15 ml)	
Kloride	10% (20 mEq/15 ml)	
Klorvess	10% (20 mEq/15 ml)	
Pfi Klor	10% (20 mEq/15 ml)	
Rum-K	15% (30 mEq/15 ml)	
Kaon-Cl 20%	20% (40 mEq/ml)	
Klor-Con	20% (40 mEq/ml)	
Kaon	20 mEq/15 ml as gluconate	• Give potassium in 2 to 4 doses per day over several days to avoid severe hyperkalemia. Give it with or after meals with a full glass of water or fruit juice to minimize GI irritation. Follow the manufacturer's recommendations for dilution.
Potassium Rougier**	20 mEq/15 ml as gluconate	
Twin-K	20 mEq gluconate and citrate	
Duo-K	20 mEq potassium and 3.4 mEq chloride per 15 ml	
Kolyum	20 mEq potassium and 3.4 mEq chloride per 15 ml	
		• Tell patients to sip liquid potassium products slowly to minimize GI irritation.
		• Don't give to patients receiving potassium-sparing diuretics.

Continued

Unmarked trade names available in the United States only.
* Also available in Canada.
** Available in Canada only.
◊ Potassium chloride unless specified.

Some Potassium Supplements
Continued

PRODUCT	AMOUNT OF POTASSIUM SUPPLIED ◊	NURSING TIPS
Powders		
K-Lor	15 mEq/packet	• Make sure powders are *completely* dissolved.
KATO	20 mEq/packet	
Kay Ciel*	20 mEq/packet	
K-Lor	20 mEq/packet	
K-Lyte/Cl*	25 mEq/packet	• A helpful tip: If patient's diet allows, mix total daily dosage of potassium powder in boiling water and then add one packet of gelatin dessert, adding usual amount of cold water to the gelatin. Once the mixture sets, it can be divided into four servings or "doses."
Kolyum	20 mEq potassium and 3.4 mEq chloride per packet (gluconate and chloride)	
Potage	20 mEq per 5-g packet	• Reconstitute Potage to make flavored soup.
Effervescent Tablets		
Kaochlor-Eff	20 mEq potassium and chloride (from potassium chloride, citrate, and bicarbonate and betaine HCL)	• Tell the patient to drink the solution after effervescence has subsided to minimize ingestion of HCO_3^-. Make sure the patient drinks all of the solution.

Continued

Unmarked trade names available in the United States only.
* Also available in Canada.
** Available in Canada only.
◊ Potassium chloride unless specified.

Some Potassium Supplements
Continued

PRODUCT	AMOUNT OF POTASSIUM SUPPLIED ◇	NURSING TIPS
Effervescent Tablets		
Klorvess	20 mEq potassium and chloride (from potassium chloride and bicarbonate and l-lysine monohydrochloride)	• Tell the patient to drink the solution after effervescence has subsided to minimize ingestion of HCO_3^-. Make sure the patient drinks all of the solution.
K-Lyte	25 mEq as bicarbonate and citrate	
Potassium-Sandoz**	12 mEq potassium and 8 mEq of chloride (from chloride and bicarbonate)	
Tablets/Capsule		
Potassium chloride	Enteric-coated 300 mg (4 mEq); 650 mg (8.7 mEq); 1 g (13.4 mEq)	• Wax matrix tablets can lodge in the esophagus and cause ulcerations in cardiac patients who have esophageal compression. In patients with esophageal stasis or obstruction, use liquid form.
K Tab	750 mg (10 mEq) potassium chloride in timed-release tablet	
Slow-K*	Sugar-coated: 600 mg (8 mEq) in wax matrix	

Continued

Unmarked trade names available in the United States only.
* Also available in Canada.
** Available in Canada only.
◇ Potassium chloride unless specified.

Some Potassium Supplements
Continued

PRODUCT	AMOUNT OF POTASSIUM SUPPLIED◇	NURSING TIPS
Tablets/Capsule *Continued*		
Kaon-Cl	Sugar-coated: 500 mg (6.67 mEq) in wax matrix	• Not recommended due to problems with GI bleeding and small bowel ulcerations.
Kaon	Sugar-coated: 5 mEq potassium gluconate	• See ***Tablets/Capsule*** on page 77
Micro K Extencaps	8 mEq per timed-release capsule	

Parenteral (available in ampules, syringes, and vials)

Potassium chloride	10 mEq/10 ml	• Always administer slowly as *dilute solutions* (see package insert).
	10 mEq/15 ml	
	20 mEq/10 ml*	
	20 mEq/20 ml	• Monitor EKG and tissue and plasma potassium levels.
	30 mEq/10 ml	
	30 mEq/12.5 ml	
	30 mEq/15 ml	• Don't administer to any patient with acute dehydration or severe renal impairment.
	30 mEq/20 ml	
	40 mEq/12.5 ml	
	40 mEq/20 ml*	
	60 mEq/30 ml	
	90 mEq/30 ml	
Potassium acetate (ampules or vials)	40 mEq/20 ml	
	50 mEq/20 ml	
	90 mEq/30 ml	

Unmarked trade names available in the United States only.
* Also available in Canada.
** Available in Canada only.
◇ Potassium chloride unless specified.

Dietary Sources of Potassium

SOURCE	SERVING SIZE	mEq
Meats		
Beef	4 oz (112 g)	11.2
Chicken	4 oz (112 g)	12.0
Scallops	5 large	30.0
Veal	4 oz (112 g)	15.2
Vegetables		
Artichokes	1 large bud	7.7
Asparagus, fresh, frozen, cooked	½ cup	5.5
raw	6 spears	7.7
Beans, dried, cooked	½ cup	10.0
Beans, lima	½ cup	9.5
Broccoli, cooked	½ cup	7.0
Carrots, cooked	½ cup	5.7
raw	1 large	8.8
Mushrooms, raw	4 large	10.6
Potato, baked	1 small	15.4
Spinach, fresh, cooked	½ cup	8.5
Squash, winter, baked	½ cup	12.0
Tomato, raw	1 medium	10.4

Continued

Dietary Sources of Potassium
Continued

SOURCE	SERVING SIZE	mEq
Fruits		
Apricots, dried	4 halves	5.0
fresh	3 small	8.0
Banana	1 medium	12.8
Cantaloupe	½ small	13.0
Figs, dried	7 small	17.5
Peach, fresh	1 medium	6.2
Pear, fresh	1 medium	6.2
Beverages		
Apricot nectar	1 cup (240 ml)	9.0
Grapefruit juice	1 cup (240 ml)	8.2
Milk, whole, skim	1 cup (240 ml)	11.4
Orange juice	1 cup (240 ml)	9.0
Pineapple juice	1 cup (240 ml)	14.4
Prune juice	1 cup (240 ml)	11.6
Tomato juice	1 cup (240 ml)	8.8

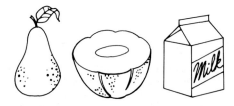

Normal Serum and Urine Calcium Values

Normal serum calcium values
Normally, serum calcium levels range from 8.9 to 10.1 mg/dl (atomic absorption), or from 4.5 to 5.5 mEq/liter. In children, serum calcium levels are higher than in adults. Calcium levels can rise as high as 12 mg/dl, or 6 mEq/liter, during phases of rapid bone growth.

Normal urine calcium values
Normal values depend on dietary intake. Males excrete < 275 mg of calcium/24 hours; females, < 250 mg/24 hours. Normal excretion of phosphate is < 1,000 mg/24 hours.

Why Is Calcium Important?

The human body contains about 1,200 grams of calcium, of which about 99% is tied up in bone. This bone calcium exists mainly as insoluble crystals, which give bone its hardness and durability. Bone calcium is physiologically inactive. Only the remaining 1% of the total body calcium found in the soft tissues and serum is the active amount we deal with in patients with calcium imbalance.

Three things influence the serum calcium balance: the deposition and resorption of bone; the absorption of calcium from the gastrointestinal tract; and the excretion of calcium in the urine and feces.

The *absorption of calcium* from the gastrointestinal tract depends mainly, of course, on the dietary intake. Dietary vitamin D promotes calcium absorption through the intestine; phosphates inhibit it. Phosphates, vitamin D, and parathyroid hormone regulate calcium excretion in the urine and feces. Parathyroid hormone, vitamin D, and phosphates all work together to both increase and decrease the serum calcium. When these elements are normal, calcium excretion remains fairly constant.

Calcium Absorption Physiology

Since most calcium compounds are insoluble, they're poorly absorbed from the intestinal tract. Vitamin D, particularly vitamin D_3, and parathyroid hormone play an important role in the intestinal absorption of calcium.

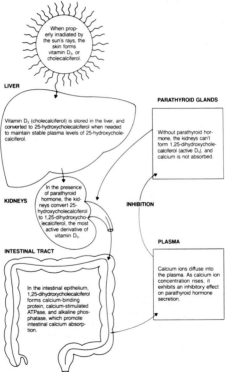

When properly irradiated by the sun's rays, the skin forms vitamin D_3, or cholecalciferol.

LIVER

Vitamin D_3 (cholecalciferol) is stored in the liver, and converted to 25-hydroxycholecalciferol when needed to maintain stable plasma levels of 25-hydroxycholecalciferol.

PARATHYROID GLANDS

Without parathyroid hormone, the kidneys can't form 1,25-dihydroxycholecalciferol (active D_3), and calcium is not absorbed.

KIDNEYS

In the presence of parathyroid hormone, the kidneys convert 25-hydroxycholecalciferol to 1,25-dihydroxycholecalciferol, the most active derivative of vitamin D_3.

INHIBITION

INTESTINAL TRACT

In the intestinal epithelium, 1,25-dihydroxycholecalciferol forms calcium-binding protein, calcium-stimulated ATPase, and alkaline phosphatase, which promote intestinal calcium absorption.

PLASMA

Calcium ions diffuse into the plasma. As calcium ion concentration rises, it exhibits an inhibitory effect on parathyroid hormone secretion.

Hypocalcemia

HYPOCALCEMIA (serum calcium below 8.5 mg/dl)

Causes
- Hypoparathyroidism
- Chronic renal failure
- Inadquate vitamin D and calcium intake
- Chronic malabsorption syndrome
- Cancer
- Hyperphosphatemia
- Hypomagnesemia
- Cushing's syndrome
- Acute pancreatitis

Signs and symtpoms
- Muscle tremors, muscle cramps, tetany, tonic/clonic seizures, paresthesia
- Alteration in normal blood clotting mechanisms, causing bleeding
- Dysrhythmias, hypotension, lengthened QT interval with normal T wave on EKG waveform
- Anxiety, irritability, twitching, Chvostek's sign, Trousseau's sign

Nursing considerations
- Place patient on cardiac monitor, as ordered. Observe him for changes in heart rate, rhythm, and EKG pattern.
- Administer calcium gluconate or calcium chloride 10% I.V., as ordered.
- Provide calcium in diet, as ordered.
- Administer vitamin D if deficiency is present, as ordered.
- Monitor serum calcium levels every 12 to 24 hours. Report a serum calcium of less than 8 mEq/liter.
- Monitor blood's prothrombin time and platelet levels.
- Administer any ordered antacids with caution; some contain phosphorus.

Special Consideration

Watch for hypocalcemia especially in patients recovering from surgery involving the parathyroids or thyroid. In such patients, diminished parathyroid hormone upsets the calcium balance. The serum calcium may drop precipitously, even resulting in tetany or seizures. For this reason, you'll want to keep calcium gluconate at the bedside after such surgery.

Hypercalcemia

HYPERCALCEMIA (serum calcium above 5.5 mEq/liter)

Causes
- Long-term immobilization (causes calcium displacement from bone to blood)
- Hyperparathyroidism
- Hypophosphatemia
- Metastatic carcinoma
- Alkalosis
- Thyrotoxicosis
- Vitamin D toxicosis
- Prolonged thiazide diuretic therapy
- Addison's disease

Signs and symptoms
- Drowsiness, lethargy, headaches, depression, apathy, irritability, confusion, personality change
- Increased incidence of kidney stones, with associated flank pain
- Muscular flaccidity
- Nausea, vomiting, anorexia, or constipation
- Polydipsia
- Polyuria
- Hypertension; enhanced effectiveness of digitalis, to the point of toxicity (administration of digitalis may cause dysrhythmias); shortening of QT interval of EKG waveform; decreased effectiveness of cardiac contractions, leading to cardiac arrest
- Pathologic fractures

Nursing considerations
- Place patient on cardiac monitor. Observe him for changes in heart rate, rhythm, and EKG pattern.
- Monitor serum calcium level frequently. Watch for cardiac dysrhythmias if serum calcium level exceeds 5.5 mEq/liter.
- If patient's on digitalis, check his digitalis serum level before administering each daily dose. Assess him for signs of digitalis toxicosis, such as vomiting, headache, fatigue, and dysrhythmias.
- Administer the following medication, as ordered: loop diuretics (never thiazide diuretics) and fluid therapy, to enhance renal calcium excretion; mithramycin (Mithracin); corticosteroids; phosphate binders.
- Administer any ordered antacids with caution; some contain calcium.
- Check urine for renal calculi and acidity.

Nursing Tip

Be aware that because over half of a patient's serum calcium is chemically inactive, he can be hypercalcemic and still have a normal lab value for serum calcium.

Checking for Trousseau's and Chvostek's Signs

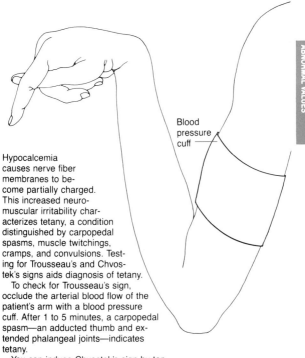

Hypocalcemia causes nerve fiber membranes to become partially charged. This increased neuromuscular irritability characterizes tetany, a condition distinguished by carpopedal spasms, muscle twitchings, cramps, and convulsions. Testing for Trousseau's and Chvostek's signs aids diagnosis of tetany.

To check for Trousseau's sign, occlude the arterial blood flow of the patient's arm with a blood pressure cuff. After 1 to 5 minutes, a carpopedal spasm—an adducted thumb and extended phalangeal joints—indicates tetany.

You can induce Chvostek's sign by tapping the patient's facial nerve adjacent to the ear. A brief contraction of the upper lip, nose, or side of the face is a positive sign.

Some Commonly Used Calcium Supplements

PRODUCT	AMOUNT OF CALCIUM SUPPLIED*	NURSING TIPS
Tablets		
Calcium gluconate ◇	27 mg/300 mg 45 mg/500 mg 54 mg/600 mg 90 mg/1 g	• Well tolerated and inexpensive. Give with meals (except dairy products) because phosphorus and corticosteroids may interfere with calcium absorption. Calcium may reduce the blood levels of tetracycline; don't give within 1 hour of each other. • Watch for symptoms of hypercalcemia. • Monitor serum calcium levels.
Calcium lactate ◇	39 gm/300 mg 78 mg/600 mg 84 mg/650 mg	
Powders		
Calcium carbonate	400 mg/g	• Less soluble than other oral forms
Calcium phosphate	230 mg/g	
Parenterals		
Calcium chloride	10% solution/10 ml (270 mg)	• Very irritating. Avoid extravasation when you give I.V. (in cardiac arrest). Not suitable for oral use.
Calcium gluconate ◇	10% solution/10 ml (90 mg)	

* 1 mEq elemental calcium equals 20 mg.
 Unmarked trade names available in the United States only.
◇ Also available in Canada.

Disorders That Affect Urine Calcium and Urine Phosphorus Levels

DISORDER	URINE CALCIUM LEVEL	URINE PHOSPHATE LEVEL
Hyperparathyroidism	Elevated	Elevated
Vitamin D intoxication	Elevated	Suppressed
Metastatis carcinoma	Elevated	Normal
Sarcoidosis	Elevated	Suppressed
Renal tubular acidosis	Elevated	Elevated
Multiple myeloma	Elevated or normal	Elevated or normal
Paget's disease	Normal	Normal
Milk-alkali syndrome	Suppressed or normal	Suppressed or normal
Hypoparathyroidism	Suppressed	Suppressed
Acute nephrosis	Suppressed	Suppressed or normal
Chronic nephrosis	Suppressed	Suppressed
Acute nephritis	Suppressed	Suppressed
Renal insufficiency	Suppressed	Suppressed
Osteomalacia	Suppressed	Suppressed
Steatorrhea	Suppressed	Suppressed

Additional Causes of Calcium Imbalance

HYPOCALCEMIA

- Vitamin D insufficiency
- Small-bowel disease
- Inadequate total parenteral nutrition
- Starvation
- Diarrhea
- Certain malignancies in which bone metastases stimulate a marked increase in osteoblastic activity
- Drugs, such as Dilantin
- Sepsis
- Burns
- Massive blood transfusions
- Markedly increased serum phosphate levels, such as in uremia, after chemotherapy for leukemia or after phosphate infusion
- Autoimmune disease
- Acute pancreatitis
- Severe liver disease
- Calcium malabsorption
- Hypomagnesemia
- Postoperative abdominal or neck surgery

HYPERCALCEMIA

- Vitamin D excess
- Milk-alkali syndrome (Burnett's syndrome)
- Malignancy with metastasis
- Malignancy without metastasis
- Sarcoidosis
- Prolonged thiazide therapy
- Immobilization
- Acute renal failure (diuretic phase)
- Hyperthyroidism
- Addison's disease
- Idiopathic hypercalcemia in infancy

Special Consideration

- Expect fairly constant calcium excretion when phosphates, vitamin D, and parathyroid hormones are normal. Suspect hypocalcemia in a patient with perioral paresthesias; twitching; carpopedal spasm; tetany; seizures; and possibly cardiac dysrhythmias.

Magnesium: Normal Serum and Urine Values

Normally, serum magnesium levels range from 1.7 to 2.1 mg/dl (atomic absorption), or from 1.5 to 2.5 mEq/liter.

Normal urinary excretion of magnesium is less than 150 mg/24 hours (atomic absorption).

Why Is Magnesium Important?

Magnesium helps cell metabolism, activates many enzyme systems, and influences the metabolism of nucleic acids and proteins. Magnesium also affects skeletal muscle directly by depressing acetylcholine release at the synaptic junction; it facilitates transportation of sodium and potassium across cell membranes (accounting for the secondary hypokalemia that occurs in hypomagnesemia); and it influences intracellular calcium levels through its effect on parathyroid hormone secretion.

An adult body contains about 2,000 mEq of magnesium. Some 60% of it is in bone, about 1% in extracellular fluid, and the rest in muscle and other soft tissues. The serum levels vary with the method of determination and may not reflect total body stores of magnesium.

A healthy person ingests about 25 mEq of magnesium daily, mostly in meat, green vegetables, whole grains, and nuts. About 10 mEq of magnesium is absorbed daily through the small bowel, and approximately the same amount is excreted daily in the urine. The rest is lost in the stool. During hypomagnesemia, however, the kidneys conserve magnesium and excrete only about 1 mEq per day.

MAGNESIUM

Hypomagnesemia

Most magnesium-related disorders result from magnesium deficiency (hypomagnesemia). This deficiency does not usually follow inadequate dietary intake alone. Such deficiency occurs only when something impairs the absorption of magnesium, or when excretion becomes too rapid. Expect to find magnesium excess in patients unable to excrete magnesium (as in chronic kidney failure) combined with excessive use of antacids, magnesium-containing cathartics, or milk of magnesia.

PROBLEM/CAUSES	SIGNS AND SYMPTOMS	NURSING CONSIDERATIONS
• Hypomagnesemia (serum magnesium level below 1.5 mEq/liter) • Starvation syndrome • Malabsorption syndrome • Postoperative complications after bowel resection • Prolonged TPN therapy without adequate magnesium • Excessive administration of mercurial diuretics • Excessive GI fluid losses from nasogastric suctioning, vomiting, diarrhea, or fistula • Hyper- and hypoparathyroidism	• Dizziness, confusion, delusions, hallucinations, convulsions • Tremors, hyperirritability, tetany, leg and foot cramps, Chvostek's sign • Dysrhythmias and vasomotor changes • Anorexia and nausea	• If patient needs antacids, give Maalox* or Mylanta* (as ordered). • Take seizure precautions, such as keeping side rails up and initiating neurochecks. • Replace magnesium losses, as ordered. Infuse magnesium replacement slowly, observing patient for bradycardia and decreased respirations. • Monitor serum magnesium levels every 6 to 12 hours during replacement therapy. Report abnormal levels immediately.

Continued

Hypermagnesemia
Continued

PROBLEM/CAUSES	SIGNS AND SYMPTOMS	NURSING CONSIDERATIONS
• Hypermagnesemia (serum magnesium level above 2.5 mEq/liter) • Renal failure • Adrenal insufficiency • Excessive ingestion of magnesium (for example, in the form of antacid gels such as Maalox* and Mylanta*) • Excessive use of magnesium-containing laxatives such as Milk of Magnesia	• Drowsiness, lethargy, confusion, coma • Bradycardia, weak pulse, hypotension, prolonged QT interval on EKG waveform, heart block, cardiac arrest (with serum levels of 25 mEq/liter) • Vague neuromuscular changes (may include tremors and hyporeflexia) • Vague GI symptoms, such as nausea	• Discontinue Maalox* and Mylanta*, if patient's receiving them. • Administer calcium gluconate I.V. as ordered. Note that calcium enhances digitalis action. • In renal failure, perform dialysis, as ordered, to remove excess magnesium. • Monitor serum magnesium levels to determine effectiveness of treatment. Watch for respiratory distress if serum magnesium levels exceed 10 mEq/liter.

Special Consideration

Be aware that low calcium and potassium levels tend to coexist with low magnesium levels. Suspect hypermagnesemia in a patient with serious hypotension accompanied by a feeling of warmth and sweating.

Unmarked trade names available in the United States only
* Also available in Canada

Some Magnesium Supplements

PRODUCT	AMOUNT OF MAGNESIUM SUPPLIED	NURSING TIPS
Tablets		
Magora-Forte	30 mg	• Excessive doses may cause laxative effect; magnesium inhibits absorption of tetracycline
Mg-Plus	60 mg	
Mg + C	20 mg	
		• With prolonged use and in patients with some degree of renal impairment, watch for symptoms of hypermagnesemia
Parenteral		
Magnesium sulfate ◊	10% solution (100 mg/ml) in 10, 20 ml ampules and 20 ml disposable units	Magnesium sulfate is used as follows: • As a nutritional supplement in hyperalimentation
	25% solution (250 mg/ml) in 10 ml ampules	• As an anticonvulsant, especially in preeclampsia or eclampsia • Administered I.V. diluted to a concentration of 20% or less
	50% solution (500 mg/ml) in 2, 10, 20 ml ampules; 5, 10, 20 ml disposable units; and 30 ml vials	• Administered deep I.M. 50% solution in adults; 20% in children • Use with caution in patients receiving diuretics, corticosteroids, and CNS depressants

Unmarked trade names available in the United States only.
◊ Also available in Canada.
◊ ◊ Available in Canada only.

Chloride: Normal Serum Values

Normally, serum chloride levels range from 100 to 108 mEq/liter. Several different assay methods can determine serum chloride levels, so individual labs may report slightly different normal values. Know what values your lab considers normal. Remember that blood samples should be processed promptly to prevent a chloride shift from plasma to red cells, causing falsely low values. Also, lab values for serum chloride may be falsely high or low if the patient is using certain drugs.

Expect *high* chloride values if the patient's taking any of the following drugs: ammonium chloride; boric acid (toxicity); ion-exchange resins; cholestyramine (Questran); oxyphenbutazone (Tandearil); phenylbutazone (Butazolidin); and excess intravenous sodium chloride.

Expect *low* chloride values if the patient is taking any of the following drugs: bicarbonate; ethacrynic acid (Edecrin); furosemide (Lasix); thiazide diuretics; or 5% dextrose in water for prolonged I.V. use (dilutional effect).

Phosphate: Normal Serum Values

Normally, serum phosphate levels range from 2.5 to 4.5 mg/dl (atomic absorption), or from 1.8 to 2.6 mEq/liter. Children have higher serum phosphate levels than adults. Phosphate levels can rise as high as 7 mg/dl, or 4.1 mEq/liter, during periods of increased bone growth.

Copper: Normal Urine Values

Normal urinary excretion of copper is 15 to 60 mcg/24 hours.

Trace Elements

More than 50 elements occur in minute quantities in human tissues. These so-called trace elements influence many physiologic processes, including enzyme function and protein and nucleic acid synthesis.

So far, only a handful of trace elements are known to cause clinical deficiency states:

- *Cobalt* is an essential component of vitamin B_{12}. Its deficiency can cause pernicious anemia.
- *Copper* deficiency has been implicated in an anemia that mimics iron-deficiency anemia. Copper-free diets have led to leukopenia and neutropenia.
- *Iodine* is essential for thyroid function. Iodine deficiency leads to hypothyroidism.
- *Zinc* appears to be necessary for wound healing and enzyme activity.
- *Manganese* seems essential for skeletal growth and calcium and phosphorus metabolism.

Normal diets and most dietary supplements and vitamins with minerals easily provide the necessary trace elements. But hyperalimentation fluids do not. So consider the need for trace element supplements for patients who must receive hyperalimentation for 3 weeks or more.

Cobalt: Critical Trace Element

A trace element found mainly in the liver, cobalt is an essential component of vitamin B_{12} and therefore is a critical factor in hematopoiesis. A balanced diet supplies sufficient cobalt to maintain hematopoiesis, primarily through foods containing vitamin B_{12}. However, excessive ingestion of cobalt may have toxic effects. Toxicity has occurred, for example, in persons who consumed large quantities of beer containing cobalt as a stabilizer, resulting in congestive heart failure from cardiomyopathy. Since quantitative analysis of cobalt alone is difficult because of the minute amount found in the body, cobalt is often measured by bioassay as part of vitamin B_{12}.

The normal cobalt concentration of human plasma is about 60 to 80 pg/ml.

Assessing Wilson's Disease

Increased serum and urine copper levels may suggest Wilson's disease. Usually, the first symptoms of this rare, inherited disorder are neurologic—rigidity, tremors, incoordination, and ataxia. Later, liver insufficiency with jaundice, ascites, and cirrhosis may develop. Kayser-Fleischer ring, a rust-colored ring around the cornea caused by copper deposits, confirms Wilson's disease.

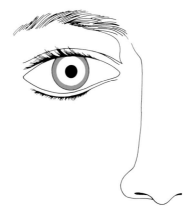

Hypophosphatemia

HYPOPHOSPHATEMIA (serum phosphate level below 1.8 mEq/liter)

Problem/Cause
- Chronic alcoholism (usually entails decreased phosphate intake)
- Prolonged phosphate-free or low-phosphate total parenteral nutrition (TPN) therapy
- Hyperparathyroidism, with resultant hypercalcemia
- Excessive use of phosphate-binding gels such as aluminum hydroxide
- Malabsorption syndrome
- Chronic diarrhea

Signs and symptoms
- Anorexia
- Mental confusion
- Muscle weakness, muscle wasting, tremors, paresthesia
- Hemolytic anemia
- Hypoxia with peripheral cyanosis
- Hypercalcemia

Nursing considerations
- Monitor calcium, magnesium, and phosphorus levels, and report any changes immediately.
- Administer potassium phosphate I.V., as ordered.
- Provide phosphate in diet, or give oral phosphate supplements, as ordered. Monitor patient for signs of hypocalcemia when giving supplements.
- If patient's receiving phosphate-binding gels such as aluminum hydroxide (Amphojel*), discontinue their use.

*Available in both the United States and Canada.

Special Consideration

Calcium and phosphorus are closely related, usually reacting together to form insoluble calcium phosphate. To prevent formation of a precipitate in the blood, calcium levels vary inversely with phosphorus; as serum calcium levels rise, phosphorus levels should decrease through renal excretion. Since the body excretes calcium daily, regular ingestion of calcium in food (at least 1 g/day) is necessary for normal calcium balance.

Hyperphosphatemia

HYPERPHOSPHATEMIA (serum phosphate level above 4.5 mg/dl)

Problem/Cause
- Excessive use of phosphate-containing laxatives or enemas
- Acute and chronic renal failure
- Excessive I.V. or oral phosphate therapy
- Cytotoxic agents
- Vitamin D toxicosis
- Hypocalcemia
- Hypoparathyroidism

Signs and symptoms
- Usually asymptomatic
- Possible metastatic calcifications
- With hypocalcemia, neuromuscular changes including cramps, tetany, or seizures

Nursing considerations
- Administer phosphate-binding gels such as aluminum hydroxide (Amphojel*), as ordered.
- Monitor serum calcium, magnesium, and phosphorus levels. Report any changes immediately.
- Observe patient for signs of hypocalcemia, such as muscle twitching and tetany.
- Discontinue antacids, such as Maalox* and Mylanta*.

*Available in both the United States and Canada.

Factors Which Increase Serum Phosphate Levels

- Excessive vitamin D intake and drug therapy with anabolic steroids and androgens may elevate serum phosphate levels.
- Improper handling of the sample, resulting in hemolysis, falsely increases serum phosphate levels.

Some Commonly Used Chloride Supplements

PRODUCT	AMOUNT OF CHLORIDE SUPPLIED	NURSING TIPS
Tablets		
Ammonium chloride	300 mg 500 mg 1 g (enteric-coated)	• Large doses may cause metabolic acidosis due to hyperchloremia. May cause gastric irritation; absorption of enteric tablets is unpredictable; administer p.c.
Syrup		
Ammonium chloride	500 mg/5ml	
Powder		
L-lysine monohydrochloride	Not available commercially	
Parenteral		
Arginine hydrochloride (R-Gene)	10% solution 300 ml 142.5 mEq/liter	• Rapid infusion may produce vomiting, irritation at infusion site, increased BUN, and severe hyperkalemia (in patients with renal disease)
Hydrochloric acid	Not available commercially	• Pharmacy usually prepares a 0.1 normal HCl solution. Use extreme caution in transcribing the doctor's order. Errors have been reported in transcribing 100 mg HCl as 100 mEq of KCl
Ammonium chloride ◊	2.14% 400 mEq/liter	• Must be administered very slowly to avoid ammonium toxicity. Don't administer I.M. or S.C.

Unmarked trade names available in the United States only.
◊ Also available in Canada.

Capillary Pressure Effects

The pressure at the arterial end of the capillary is almost three times higher than at the venous end. This pressure differential is an important key to the movement of fluids through the capillary network.

Fluid leaves the capillary bed at the arterial end. It reenters the capillary bed at the venous end through reabsorption. Because the net outward force is greater than the net inward force, 10% of the fluid filtered at the arterial end is not reabsorbed at the venous end. If this continued unchecked, the interstitial space would eventually be overloaded with fluid.

But the lymph system picks up this remaining 10% along with certain proteins and other large molecules too big to reenter the capillary and returns them to the heart.

Under Pressure

Four different pressures move fluid through the capillary membrane:

Capillary pressure is the pressure that fluid within the capillary exerts outward against the membrane, like water filling a balloon.

Interstitial fluid pressure exerts a similar pressure inward against the capillary membrane.

Plasma colloid osmotic pressure, a magnetlike attraction of proteins, pulls fluid from the interstitial space into the capillary.

Interstitial fluid colloid osmotic pressure, another "protein magnet" in the opposite direction, draws fluid from the capillary out into the third space.

Evaluating Edema

Pitting edema is evaluated on a four-point scale: from +1 (a barely detectable pit, as in Figure 1) to +4 (a deep and persistent pit, approximately 1″ (2.54 cm) deep, as in Figure 2). An adult patient can accumulate up to 10 lb (4.5 kg) of fluid before you will be able to detect a pit. The skin pits against a bony surface, such as the subcutaneous aspect of the tibia, fibula, sacrum, or sternum.

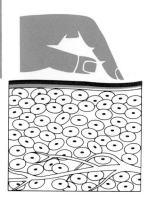

1. SLIGHT, +1 PITTING EDEMA

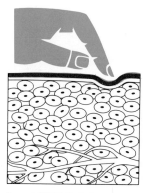

2. +4 PITTING EDEMA

Continued

Edema can become so severe that pitting is not possible; the tissue becomes so full that fluid can't be displaced. Subcutaneous tissue becomes fibrotic, and, as a result, the surface tissue feels rock-hard. Over time, this condition can develop into brawny edema (Figure 3). For example, a mastectomy patient whose axillary nodes have been removed may develop brawny edema in her affected arm. The patient's arm looks like a pig's skin and becomes hard or gelatinous to the touch. Brawny edema can also indicate lymphatic obstruction.

Protect all edematous extremities from injury: A large quantity of edema makes the skin prone to sloughing and ulceration.

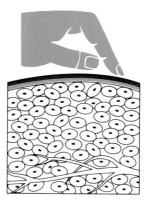

3. BRAWNY EDEMA

Nursing Care of Edematous Extremities

- Keep skin clean and dry
- Protect from injury
—while bathing or handling for any reason, do so gently and carefully
—remove room equipment, such as footstools, that could easily bump the edematous part
—avoid constrictive clothing (garters, elastic stockings)
- Keep edematous part elevated
- Observe for and record
—sores and blisters
—increasing size
—absence of pulses
—numbness
—tingling
—coldness
—cyanosis

Special Consideration

- Look for edema in patients with excessive reabsorption of fluid (heart failure, cirrhosis), inadequate elimination from failing renal function (nephrotic syndrome, acute glomerulonephritis, acute tubular necrosis, chronic renal disease), or congestive heart failure.
- Suspect edema if your patient has these signs and symptoms: excessive weight gain, elevated blood pressure, dyspnea, and neck vein distention.
- If your patient develops pulmonary edema, place him in a sitting position and administer oxygen, as ordered; give, as prescribed, digitalis, morphine I.M., and fast-acting diuretics, such as furosemide.

Assessing Fluid Excess

BODY SYSTEM	FLUID EXCESS
Cardiovascular	• Bounding pulse • Hypertension
Gastrointestinal	• Anorexia • Nausea and vomiting • Acute weight gain (greater than 5%) • Increased abdominal girth
Integumentary	• Pitting edema • Pallor • Facial edema
Musculoskeletal	• Muscle hypertonicity
Neurologic	• Confusion • Apathy
Ophthalmic	• Periorbital edema
Renal	• Polyuria (if kidneys are healthy)
Respiratory	• Shortness of breath • Hoarseness • Productive cough • Moist crackles • Dyspnea

Overhydration

PROBLEM/ CAUSE	SIGNS AND SYMPTOMS	NURSING CONSIDERATIONS
• Increased water ingestion, including psychogenic water ingestion	• Weight gain	• For prevention, exercise caution when administering hypotonic solutions such as 0.45% NaCl.
• Excessive ADH secretion	• Muscle weakness	
• Oliguric phase of renal disease	• Moist crackles; dyspnea	• For prevention, administer fluids cautiously (at specified rates), especially in patients with cardiopulmonary disease.
• Excessive administration of I.V. fluids	• Increased blood pressure	
	• Lethargy, apathy, disorientation	
	• Possible reduced hematocrit	
	• Possible reduced serum sodium level (from dilution)	
	• Abdominal cramping	

Dehydration

PROBLEM/CAUSE	SIGNS & SYMPTOMS	NURSING CONSIDERATIONS
• Decreased water intake from dysfunctional thirst mechanism • Antidiuretic hormone (ADH) insufficiency • Diuretic phase of renal disease • Excessive use of diuretics • Administration of osmotic agents (such as mannitol or dextrans), which may increase diuresis • Hyperglycemia (in uncontrolled diabetes) leading to osmotic diuresis • Fever • Excessive diaphoresis • Excessive gastrointestinal fluid losses from nasogastric drainage, diarrhea, or vomiting	• Low-grade fever • Flushed skin, dry mucous membranes, poor skin turgor • Hypotension • Rapid pulse • Thirst • Muscle weakness • Alteration in urine output (usually oliguria) • Elevated hematocrit and serum sodium levels • Weight loss • Lethargy, disorientation, coma (in severe dehydration)	• For prevention, administer any hypertonic solutions cautiously, staying alert for signs of dehydration. • Observe patient closely for insensible water loss from hyperventilation and excessive diaphoresis. • Administer fluids, as ordered.

Nursing Tip

Look for signs and symptoms of extreme fluid loss in ill or severely debilitated patients, in those with burns and massive injuries, and in those who can't ask for or obtain water, are comatose or unable to swallow, have a draining wound or a nasogastric tube in place, or are being suctioned frequently.

Assessing Fluid

BODY SYSTEM	FLUID LOSS/ELECTROLYTE IMBALANCE
Cardiovascular	Elevated temperature; rapid, weak pulse; extremities cold to touch; hypotension
Gastrointestinal	Anorexia, nausea and vomiting, thirst, acute weight loss (greater than 5%), constipation, abdominal cramps, longitudinal wrinkles on tongue
Integumentary	Poor skin turgor, dry mucous membranes, drawn facial expression
Musculoskeletal	Muscle weakness
Neurologic	Lethargy, indifference, confusion, coma
Ophthalmic	Sunken eyes
Renal	Oliguria, anuria
Respiratory	Shortness of breath; deep, rapid breathing

Nursing Tip

Watch for symptoms that predispose a patient to extreme fluid loss, such as fever, fluid shifts, vomiting, hemorrhage, and hyperventilation.

DEHYDRATION

The Pinch Test

If you pinch the skin on a patient's forearm or sternum, under normal circumstances the skin will resume its shape quickly—within a few seconds. If the skin remains wrinkled for 20 to 30 seconds, however, the patient has poor skin turgor. Reduced or poor skin turgor may indicate dehydration, rapid weight reduction, or senile cutaneous atrophy. Report your findings to the doctor.

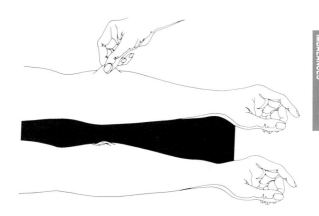

Diuretics

Diuretics reduce the body's total volume of water and salt by increasing their urinary excretion. This occurs mainly because diuretics impair sodium chloride reabsorption in the renal tubules. Diuretics can be classified according to chemical structure (thiazide and thiazide-like drugs), location of salt- and water-depleting effect on the kidneys' nephrons (loop diuretics), and pharmacologic activity (carbonic anhydrase inhibitors and miscellaneous drugs with other mechanisms).

Thiazide diuretics	benzthiazide chlorothiazide cyclothiazide hydrochlorothiazide hydroflumethiazide methyclothiazide polythiazide trichlormethiazide
Thiazide-like diuretics	chlorthalidone metolazone quinethazone
Loop diuretics	bumetanide ethacrynate sodium ethacrynic acid furosemide
Carbonic anhydrase inhibitors	acetazolamide acetazolamide sodium dichlorphenamide methazolamide
Miscellaneous diuretics	amiloride hydrochloride indapamide mannitol spironolactone triamterene urea (carbamide)

Comparing Thiazide and Thiazide-like Diuretics

Many thiazides have been developed since chlorothiazide (Diuril) was introduced in 1958. Chemical derivatives of the thiazides are known as thiazide-like diuretics. Closely related pharmacologically, thiazides and thiazide-like diuretics are used to treat edema and essential hypertension.

All have similar mechanisms of action and side effects, as well as similar therapeutic effects, under most clinical conditions. Their durations of action, however, may differ. Those that act for 24 hours can be given once daily, promoting patient compliance.

DIURETIC	EQUIVALENT DOSE (mg)	MAXIMUM DAILY DOSE (mg)	DURATION OF ACTION (hr)
Thiazide			
bendroflumethiazide	5	20	18 to 24
benzthiazide	50	200	12 to 18
chlorothiazide	500	1,000	6 to 12
cyclothiazide	2	6	24 to 36
hydrochlorothiazide	50	200	12
hydroflumethiazide	50	200	10 to 12
methyclothiazide	5	10	24
polythiazide	2	4	24 to 36
trichlormethiazide	2	4	24
Thiazide-like			
chlorthalidone	50	200	46 to 72
metolazone	5	20	12 to 24
quinethazone	50	200	18 to 24

Major Uses of Diuretics

TREATMENT	THIAZIDE DIURETICS	THIAZIDE-LIKE DIURETICS	LOOP DIURETICS
Treatment of essential hypertension	•	•	•
Treatment of edema associated with congestive heart failure, cirrhosis of the liver, and renal disease	•	•	•
Treatment of pulmonary edema			•
Reduction of intracranial pressure in hydrocephalus			
Reduction of intraocular pressure			
Treatment of open-angle glaucoma (as adjunct)			
Prophylaxis for epilepsy (as adjunct)			
Treatment of primary aldosteronism			
Potentiation of effects of mercurial diuretics			

DEHYDRATION

	CARBONIC ANHYDRASE INHIBITORS	INDAPAMIDE	AMILORIDE	MANNITOL	SPIRONOLACTONE	TRIAMTERENE	UREA
		•	•		•	•	
		•	•		•	•	
	•			•			•
	•			•			•
	•						
	•						
					•		
	•						

Understanding ABGs

The best diagnostic tool in a respiratory emergency, ABG analysis provides vital information about your patient's respiratory function within minutes. Based on blood specimens obtained from an arterial line or by percutaneous arterial puncture, this diagnostic test determines alveolar ventilation by measuring how much oxygen the lungs deliver to the blood and how efficiently

	NORMAL	RESPIRATORY ACIDOSIS
Possible causes		Impaired alveolar ventilation, respiratory depressants, intracranial tumors
Symptoms		Lethargy, shallow, irregular respirations, disorientation
Signs		Hypoventilation, asterixis
pH	7.35 to 7.45	Decreased
PaO_2	90 to 95 mm Hg	Normal or decreased
$PaCO_2$	34 to 46 mm Hg	Increased
HCO_3	24 to 26 mEq/liter	Normal (uncompensated)

ACID-BASE IMBALANCES

the lungs eliminate carbon dioxide. It also measures blood pH, which may reveal acidemia or alkalemia. Comparing pH with $PaCO_2$, you can determine if your patient's acidemia or alkalemia stems from impaired gas exchange or a metabolic disorder.

RESPIRATORY ALKALOSIS	METABOLIC ACIDOSIS	METABOLIC ALKALOSIS
Ventilatory support, hyperventilation, CNS disease, anxiety, persistent fever, liver disease, CHF, pulmonary embolism	Salicylate intoxication, renal disease, diabetes, lactic acidosis, diarrhea, biliary fistulae	Vomiting, diuretics, hyperadrenocorticism, alkali ingestion, hyperaldosteronism, nasogastric suction
Hyperactive reflexes, blurred vision, tetany, vertigo, muscle cramps, sighing, diaphoresis	Kussmaul's respirations, restlessness, disorientation	Weakness, paralysis, leg cramps, paresthesias
Hyperventilation, latent tetany	Shock, coma, tachypnea	Hypokalemic symptoms (nausea, weakness)
Increased	Decreased	Increased
Normal or decreased	Normal or increased	Normal or decreased
Decreased	Normal (uncompensated)	Normal (uncompensated)
Normal (uncompensated)	Decreased	Increased

Interpreting ABG Measurements

Oxygenation. PaO_2 and SaO_2 values reveal whether your patient's oxygen level is adequate. If he has mild hypoxemia, his PaO_2 level reveals the condition earlier than his SaO_2 value. In fact, PaO_2 can drop to as low as 50 mm Hg with no considerable change in SaO_2. However, SaO_2 will fall suddenly when PaO_2 drops below 50 mm Hg. *Note:* Abnormally low PaO_2 and SaO_2 values don't necessarily indicate hypoxemia. For example, an infant's PaO_2 is usually between 40 to 60 mm Hg; an elderly person's is usually below 80 mm Hg.

Ventilation. Does your patient have an abnormally high $PaCO_2$ value? Chances are he's hypoventilated and retained excessive CO_2 (hypercapnia). If his $PaCO_2$ value is abnormally low, he's hyperventilated and lost excessive CO_2 (hypocapnia).

Acid-base status. To find out if your patient's acidotic or alkalotic, see the information on pages 112-113.

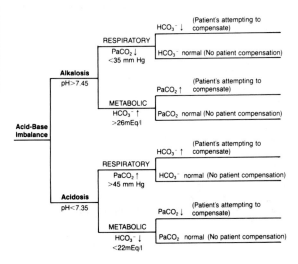

ACID-BASE IMBALANCE

Metabolic Alkalosis

Metabolic alkalosis is a state of *increased* bicarbonate (base) and *decreased* acid in the blood, resulting from conditions that cause:
- severe acid loss
- decreased serum potassium and chloride
- excessive bicarbonate intake.

Arterial pH level is above 7.45; HCO_3 is above 26 mEq/liter.

PREDISPOSING FACTORS

- Vomiting
- GI suctioning
- Diuretic therapy
- Corticosteroid therapy
- Cushing's syndrome
- Excessive bicarbonate intake
- Hypokalemia
- Hypercalcemia

SIGNS AND SYMPTOMS

- Neuromuscular irritability
- Tetany
- Twitching
- Seizures
- Central nervous system depression that may progress to coma
- Cardiac dysrhythmias
- Nausea and vomiting
- Hypoventilation (a sign that respiratory compensation's beginning)

COMPENSATION

In the presence of high HCO_3^-, the respiratory system compensates with *hypoventilation* to *increase* H_2CO_3 (as reflected in $PaCO_2$) and to bring pH to normal by adjusting the ratio of HCO_3^- to H_2CO_3 to 20:1 (normal).

INTERVENTIONS

- Give the patient normal saline solution and potassium I.V.
- Evaluate and correct his electrolyte imbalances.
- If his alkalosis is severe, give ammonium chloride I.V.
- Observe precautions to prevent seizures.
- Monitor his vital signs and fluid balance.
- Discontinue diuretics, if previously given.
- Treat the underlying cause, as ordered.

Metabolic Acidosis

Metabolic acidosis is a state of *excess* acid accumulation and *deficient* bicarbonate (base) in the blood, resulting from conditions that cause:
- excessive fat metabolism in the absence of carbohydrates
- anaerobic metabolism
- underexcretion of metabolized acids or inability to conserve base
- loss of sodium bicarbonate from the intestines.

Arterial pH level is below 7.35; HCO_3 is below 22 mEq/liter.

PREDISPOSING FACTORS

- Diabetes
- Addison's disease
- Renal failure
- Starvation
- Ethanol intoxication
- Tissue hypoxia
- Diarrhea
- Intestinal malabsorption
- Salicylate poisoning
- Low-carbohydrate, high-fat diet

SIGNS AND SYMPTOMS

- Headache and lethargy
- Central nervous system depression that may progress to coma
- Cardiac dysrhythmias
- Nausea and vomiting
- Anorexia
- Dehydration
- Kussmaul's respirations (a sign that respiratory compensation's beginning)

COMPENSATION

In the presence of low HCO_3^-, the respiratory system compensates with *hyperventilation* to *decrease* H_2CO_3 (as reflected in $Paco_2$) and to bring pH to normal by adjusting the ratio of HCO_3^- to H_2CO_3 to 20:1 (normal).

INTERVENTIONS

- Give the patient sodium bicarbonate I.V.
- Evaluate and correct his electrolyte imbalances.
- Observe precautions to prevent seizures.
- Monitor his vital signs and fluid balance.
- Treat the underlying cause, as ordered.

DIABETES MELLITUS

Diabetic Ketoacidosis

Suspect DKA in any patient who is comatose, dehydrated, and who has deep, labored respiration.

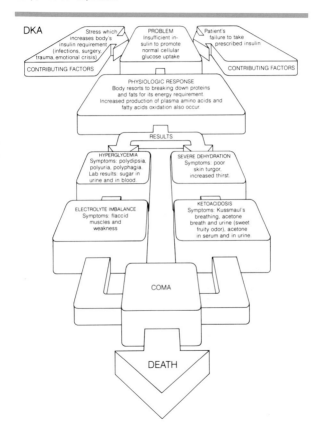

Comparing DKA, HHNC, and Hypoglycemia

Controlling blood glucose levels with diet and drugs is the key to medical management of diabetes. No less important, however, is the management of acute and chronic complications of this disease. The acute complications, if identified correctly, are reversible with treatment; the chronic complications account for most of the morbidity and mortality.

FACTOR	DIABETIC KETOACIDOSIS
Type of diabetic	Insulin-dependent (Type I)
Signs and symptoms	History of nausea, vomiting, warm and dry skin, flushed appearance, dry mucous membranes, soft eyeballs, Kussmaul's respirations or tachypnea, abdominal pain, alterations in level of consciousness, hypotension, tachycardia, acetone breath
Precipitating factor	Undiagnosed diabetes, neglect of treatment, infection, cardiovascular disorder, other physical or emotional stress

Diabetic patients are vulnerable to three types of metabolically induced coma. (The term "coma" as used here denotes a diabetic crisis accompanied by unconsciousness or altered mentation.) The three states are hypoglycemia, DKA, and HHNC.

HHNC	HYPOGLYCEMIA
Non-insulin-dependent (Type II); nondiabetic person	Insulin-dependent (Type I)
Same as in DKA except without Kussmaul's respirations or acetone breath	Nausea, hunger, malaise, cool and moist skin or diaphoresis, pallor, bradycardia, bradypnea, visual disturbances, alterations in level of consciousness; memory loss, confusion, hallucinations, generalized or focal seizures, status epilepticus, primitive movements (sucking, smacking lips, picking or grasping, Babinski's reflex) may be present
Undiagnosed diabetes, infection or other stress, drugs (phenytoin sodium, thiazide diuretics, mannitol, steroids), dialysis, I.V. hyperalimentation, acute pancreatitis, central nervous system disorders, major burns treated with high concentrations of sugar	Delayed or omitted meal, insulin overdose, excessive exercise without alterations in food or insulin

Continued

Comparing DKA, HHNC, and Hypoglycemia
Continued

FACTOR	DIABETIC KETOACIDOSIS
Onset of symptoms	Slow (hours to days)
Laboratory findings:	
Blood glucose	Usually less than 800 mg/dl
Serum sodium	Normal or decreased
Serum potassium	Normal or elevated at first, then decreased
Blood urea nitrogen	Elevated
Serum ketones	Elevated
White blood cells	Elevated
Hematocrit	Elevated
Urine glucose	Elevated
Urine ketones	Elevated
Arterial blood gases	Metabolic acidosis with compensatory respiratory alkalosis
Treatment	Insulin; I.V. fluids, such as normal or possibly half-normal saline solution; potassium* when urine output is adequate; sodium bicarbonate if pH is less than 7.0

HHNC	HYPOGLYCEMIA
Slow (hours to days)	Rapid (minutes to hours)
Over 800 mg/dl	60 mg/100 ml or less
Elevated, normal, or low	Normal
Same as DKA	Normal
Elevated	Normal
Normal	Normal
Elevated	Normal or elevated
Elevated	Normal
Elevated	Normal
Normal	Normal
Normal or slight metabolic acidosis	Normal or slight respiratory acidosis
Insulin; I.V. fluids, such as half-normal or normal saline solution; potassium when urine output is adequate	Candy; glucose paste; orange juice if awake; dextrose 50% I.V. push; 5% to 10% dextrose in water I.V. drip; glucagon; epinephrine

*Potassium phosphate, rather than potassium chloride, may be used since patients may have hypophosphatemia.

What Happens in Unchecked Diabetic Ketoacidosis

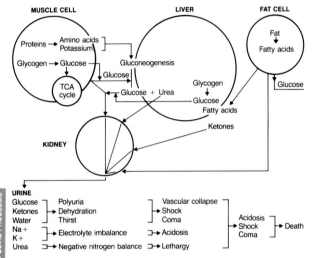

Potassium in DKA

After the second stage of treatment, potassium levels are likely to be low because:
- intravenous fluids dilute the plasma
- improved renal function after correction of fluid volume deficit promotes potassium excretion
- insulin encourages cellular uptake of potassium while glucose lowers its serum levels
- cellular glycogenesis uses potassium, further lowering serum levels
- potassium moves spontaneously into the cells (higher to lower concentration) to correct the cellular potassium deficit.

So, to prevent muscle flaccidity, rapid and shallow respirations (which replace Kussmaul's respirations), and weakness, carefully monitor potassium levels and correct any deficit, as needed. For patients who can eat or drink fluids, foods having a high potassium content or oral K+ supplements offer the easiest replacement. For those who cannot, potassium chloride may be added to the intravenous infusions. The rate of the infusion should not exceed 20 mEq/hour or 100 mEq/12 hours. Generally, the treatment goal is to replace 25% to 50% of the potassium deficit in the first 24 hours.

Never give concentrated potassium salts directly through intravenous tubing. When adding potassium to an I.V. container already hanging or lying on a counter surface, remember to mix it well in the solution. Potassium is a heavy salt and can settle near the neck of the bottle or plastic bag. Unless you disperse it thoroughly throughout the solution, the patient may get a concentrated dose of potassium with rapid fluid administration and may develop cardiac arrest.

Diabetic Ketoacidosis: Prelude to Shock?

A patient develops DKA when he has too little insulin in his body to control his blood glucose level, resulting in hyperglycemia and glycosuria.

Glucose is a heavy molecule that tends to stay in the extracellular fluid (ECF) unless pulled into a cell by insulin. When ECF glucose levels rise, glucose tends to pull water out of tissue cells. And, the patient with hyperglycemia will also experience osmotic diuresis. That's why he rapidly becomes dehydrated, setting him up for possible hypovolemic shock.

Another characteristic of DKA is ketosis. In the absence of insulin, excessive fat breakdown occurs, releasing free fatty acids. These acids then produce ketoacids in the liver. Under normal conditions, these acids are buffered and excreted. But DKA overwhelms the body's normal buffering systems, allowing metabolic acidosis to mount. Any patient with Type I diabetes is at particular risk of DKA.

Treating Diabetic Crisis: Your Role

Whether your patient has diabetic ketoacidosis (DKA) or hyperglycemic hyperosmolar nonketotic coma (HHNC), your first priorities are fluid replacement and insulin administration. But, of course, the details differ according to the diagnosis. Here's how you fit in.

What to do for DKA. Assume that your patient is dehydrated—and possibly in hypovolemic shock. Get the fluids going—stat. Follow these guidelines:
- Give fluids, as ordered.
- Assist with the collection of blood specimens, as ordered, for arterial blood gas, glucose, and electrolyte studies.
- Give insulin, as ordered, using an infusion pump to ensure dosage accuracy.

Continued

Treating Diabetic Crisis: Your Role
Continued

Chances are, the doctor will take a low-dose approach; for example, ordering 10 to 30 units of regular insulin I.V. as a loading dose followed by an infusion of 4 to 8 units of regular insulin per hour. A low-dose approach gradually reverses hyperglycemia and reduces the danger of rebound hypoglycemia, which can cause brain damage. *Important:* Insulin may have an affinity for plastic I.V. tubing. According to hospital policy, wash 50 to 100 ml of the insulin-containing solution through the tubing before starting the infusion. Then piggyback the insulin infusion into the primary I.V. line.

• Throughout treatment, regularly monitor and document vital signs, blood glucose levels, fluid intake and output, and urine glucose and acetone levels.

• As ordered, add potassium to the I.V. when sufficient renal function is established and blood glucose levels begin to fall. The osmotic diuresis characteristic of DKA may contribute to hypokalemia, which usually develops within 4 to 6 hours after onset of treatment.

• As your patient recovers, provide emotional support and reassurance. Before he's discharged from the hospital, teach him how to avoid DKA in the future.

What to do for HHNC. Treatment for HHNC is similar to that for DKA, with several variations. First, because the typical patient with HHNC is even more severely dehydrated than a patient with DKA, he needs even more fluid, perhaps several liters. *Caution:* Because a patient with HHNC is likely to be elderly, he may have a cardiac or renal condition. Closely monitor him for signs of fluid overload. Remember, fluid overload is especially dangerous for such a patient.

Second, you'll probably give a *smaller* insulin dose to a patient with HHNC than you would to a patient with DKA—even though an HHNC patient's glucose level is higher. The reason is that HHNC patients appear to be unusually sensitive to insulin.

Two Paths to Hypoglycemia

Insulin and glucagon, which originate in the islets of Langerhans in the pancreas, exert opposite physiologic effects: insulin decreases the blood glucose level and glucagon increases it. Too much insulin causes hypoglycemia; too little glucagon may cause it also.

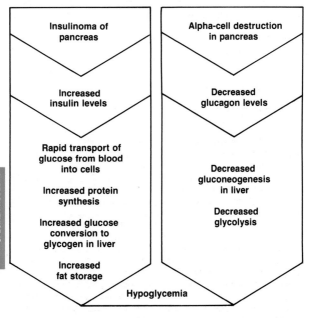

Hormones out of Balance

Normally, homeostatic mechanisms maintain blood glucose within narrow limits (60 to 120 mg/dl). The body burns available glucose and stores the rest as glycogen in the liver and muscles. When the glucose level drops, the liver converts glycogen back to glucose (glycogenolysis) or makes new glucose from noncarbohydrate sources, such as amino acids or fatty acids (gluconeogenesis). Hormones maintain the delicate balance between glucose production and use. *Insulin* prevents hyperglycemia; *epinephrine, glucogen, growth hormone,* and *cortisol* act as counterregulatory hormones to prevent hypoglycemia. But this balance is upset when your patient has either reactive or fasting hypoglycemia.

Emergency Hypoglycemia Treatment

Once you've brought your diabetic patient's hypoglycemic crisis under control, take some time to teach his family how to recognize the signs and symptoms of insulin shock (hunger, trembling, confusion) and how to intervene properly. Explain to the family that they must raise his blood glucose level *immediately* to prevent permanent brain damage and even death.

If he's conscious, someone should give him any one of the following:
• 4 oz of apple juice, orange juice, or ginger ale
• 3 oz of nondiet cola or other soda
• 2 oz of corn syrup, honey, or grape jelly
• 5 hard candies, 6 jelly beans, or 10 gumdrops.

If he's *unconscious* or if he has trouble swallowing, someone will have to administer glucagon subcutaneously.

Teach your patient's family how to prepare and administer a subcutaneous injection, and tell them to always keep a supply of glucagon at home.

Diabetes Insipidus

This disorder may occur after intracranial surgery. A patient's posterior pituitary gland temporarily stops secreting antidiuretic hormone. His renal tubules then fail to reabsorb enough water. When this happens, he'll complain of severe thirst and frequent urination. His urinary output can reach 10 liters or more a day. Its specific gravity may be as low as 1.001.

Watch for this problem, and report it at once; dehydration can occur rapidly. See that your patient gets the fluid and electrolyte replacement therapy the doctor orders.

Sometimes diabetes insipidus lasts only 2 or 3 days before correcting itself. However, if problems continue or your patient loses too much fluid, the doctor may ask you to administer posterior pituitary extract.

Treatment of Diabetes Insipidus

Until the cause of diabetes insipidus can be identified and eliminated, administration of various forms of vasopressin or of a vasopressin stimulant can control fluid balance and prevent dehydration:

- *vasopressin tannate:* an oil preparation administered I.M.; effective for 48 to 96 hours
- *vasopressin injection:* an aqueous preparation administered subcutaneously or I.M. several times a day, since effectiveness lasts only for 2 to 6 hours; often used in acute disease
- *desmopressin:* nasal spray absorbed through the mucous membranes; effective up to 20 hours
- *lypressin:* short-acting nasal spray with significant disadvantages—its variable dosage, and its adverse effects, including nasal congestion and irritation, ulcerated nasal passages (with repeated use), substernal chest tightness, coughing, and dyspnea
- *chlorpropamide:* reduces the polyuria of diabetes insipidus, possibly by releasing ADH or potentiating its effects.

Dehydration Test for Diabetes Insipidus

The dehydration test measures urine osmolality, which reflects renal concentrating capacity, after a period of dehydration and after subcutaneous injection of the pituitary hormone vasopressin. Comparison of the two osmolalities permits reliable diagnosis of diabetes insipidus, a metabolic disorder characterized by vasopressin (antidiuretic hormone) deficiency. Simply measuring urine osmolality after a period of water deprivation doesn't itself confirm vasopressin deficiency; however, subsequent injection of vasopressin raises urine osmolality beyond normal limits only in patients with diabetes insipidus.

To achieve dehydration, withhold fluids the evening before and the morning of the test. Collect a urine sample at hourly intervals in the morning for osmolality measurement. At noon, or after osmolality increases less than 30 mOsm/kg each hour for 3 consecutive hours, draw a blood sample for osmolality measurement. If serum osmolality exceeds 288 mOsm/kg, the level of adequate dehydration, inject 5 units of vasopressin subcutaneously. Within an hour, collect a urine specimen for osmolality measurement.

Clinical Alert: During dehydration, weigh the patient and monitor vital signs every 2 hours; a 1-kg weight loss normally accompanies adequate dehydration. In a patient with polyuria exceeding 10 liters/day, withhold fluids only during the morning of the test; if his weight loss exceeds 2 kg, discontinue the test.

In a patient with normal neurohypophyseal function, urine osmolality after vasopressin injection doesn't rise more than 9% of the maximum dehydration osmolality. A larger increase indicates diabetes insipidus. In a patient with polyuria caused by renal disease, potassium depletion, or nephrogenic diabetes insipidus, urine osmolality increases slightly during dehydration but not at all after vasopressin injection.

Detecting Diabetes Insipidus

The dehydration test measures urine osmolality after a period of dehydration and after subcutaneous injection of vasopressin. Plasma

PLASMA OSMOLALITY (mOsm/kg) AFTER DEHYDRATION	URINE OSMOLALITY (mOsm/kg) AFTER DEHYDRATION
288 to 291	700 to 1,400
310 to 320	100 to 200
295 to 305	250 to 500
310 to 320	100 to 200

Normal urine osmolality greater than 800 mOsm/kg water
Normal plasma (serum) osmolality 280 to 294 mOsm/kg

DIABETES INSIPIDUS

osmolality is determined just before the injection of vasopressin. Comparison of the two urine osmolalities allows diagnosis of diabetes insipidus and helps to identify its form.

CHANGE IN URINE OSMOLALITY AFTER VASOPRESSIN INJECTION	CLINICAL SIGNIFICANCE
None	Normal
100% or greater	Complete central diabetes insipidus
9% to 67%	Partial central diabetes insipidus
None	Nephrogenic diabetes insipidus

Lab Values in Primary Adrenal Insufficiency

Plasma:
- ACTH: elevated—150 mg/ml
- 17-ketosteroids: decreased—10 mg/24 hours
- Aldosterone: decreased—0.015 mcg/dl
- Sodium (Na): decreased—136 to 145 mEq/liter
- Potassium (K): elevated—3.5 to 5.0 mEq/liter
- Chloride (Cl): decreased—95 to 109 mEq/liter
- Bicarbonate (HCO_3^-): decreased—22 to 26 mEq/liter
- Sugar (BS): occasionally decreased—60 to 120 mg/dl
- Eosinophil count: increased
- Cortisol: normal or decreased (normal is 5 to 25 mcg/dl between 8 and 10 a.m., and 2 to 18 mcg/dl between 4 and 6 p.m.)

24-hour urine:
- 17-hydroxycorticosteroids: may be decreased—3.0 to 14.5 mg/24 hours in males; 3.0 to 12.9 mg/24 hours in females
- Aldosterone: decreases—2 to 26 mcg/24 hours.

Special Consideration

Remember these important points when caring for a patient at risk of thyroid or adrenal crisis:
- Prevent corneal ulceration in patients with proptosis by keeping the eyes moist at all times.
- Know that a normal cortisol level determined from random samples may not rule out adrenal insufficiency.
- Watch for adrenal insufficiency's early signs and symptoms: sweating, hunger, blurred vision, tremors, restlessness, pallor, muscle twitching, convulsions, tachycardia, weakness, or headaches.
- Teach your patient about his illness and emphasize the importance of complying with corticosteroid therapy.

What Happens in Acute Hypoparathyroidism

Hypoparathyroidism is a deficiency of parathyroid hormone (PTH). Causes of acute hypoparathyroidism include:
- injury to the parathyroid glands
- accidental removal of the parathyroid glands during thyroidectomy or other neck surgery.

Other, less common, causes include:
- autoimmune disease
- tumor
- tuberculosis
- sarcoidosis
- hemochromatosis
- severe magnesium deficiency associated with alcoholism and intestinal malabsorption (causes a reversible form of hypoparathyroidism).

Normally, PTH (with the aid of vitamin D) maintains serum calcium and phosphate levels by:
- balancing calcium levels in bone, blood, kidneys, intestines, and soft tissues
- maintaining an inverse relationship between serum calcium and serum phosphate (this means that if the level of one begins to fall, the other rises, because the product of the two is constant).

Diminished or absent PTH levels disrupt this balance, however, causing:
- excessive reabsorption of phosphate by the renal tubules
- decreased release of calcium from bone
- decreased renal calcium reabsorption
- decreased intestinal calcium absorption.

The result? Severe hypocalcemia and hyperphosphatemia, which can lead to seizures, tetany, laryngospasm, and central nervous system abnormalities.

Laboratory Findings in Endocrine Disorders

ELECTROLYTES	INCREASED LEVELS	DECREASED LEVELS
Calcium	Adrenal hypofunction Hyperparathyroidism Hyperthyroidism	Hypoparathyroidism Cushing's syndrome
Chloride	Hyperparathyroidism	Primary hyperaldosteronism Adrenal hypofunction
Magnesium	Adrenal hypofunction Early-stage diabetic acidosis	Hyperparathyroidism Hyperthyroidism Primary hyperaldosteronism
Phosphate	Acromegaly Hypoparathyroidism	Hyperparathyroidism
Potassium	Adrenal hypofunction Diabetic ketoacidosis (DKA)	Cushing's syndrome Primary hyperaldosteronism
Sodium	Cushing's syndrome Diabetes insipidus Primary hyperaldosteronism Adrenal hypofunction Chronic primary adrenocortical insufficiency	Hyperosmolar hyperglycemic nonketotic coma (HHNC)

Electrolyte Imbalances Resulting from Endocrine and Metabolic Emergencies

Remember that your patient's endocrine or metabolic emergency can cause electrolyte imbalances that are themselves actual or potential emergencies. Here's a list of some electrolyte imbalances to watch for.

DKA
- Hyponatremia
- Hyperkalemia
- Hypomagnesemia
- Hypophosphatemia
- Hypochloremia

Adrenal crisis
- Hypernatremia
- Hypercalcemia
- Hypophosphatemia
- Hyperkalemia
- Hypochloremia

Acute hypoparathyroidism
- Hypocalcemia
- Hypomagnesemia
- Hyperphosphatemia

HHNC
- Hyponatremia
- Hyperkalemia
- Hypophosphatemia
- Hypochloremia

Thyrotoxic crisis
- Hypercalcemia
- Hypomagnesemia
- Hyperphosphatemia

Myxedema crisis
- Hyponatremia
- Hypercalcemia

Blood Urea Nitrogen Values

BUN values normally range from 8 to 20 mg/dl.

Serum Creatinine Values

Creatinine concentrations in males normally range from 0.8 to 1.2 mg/dl; in females, from 0.6 to 0.9 mg/dl.

Serum Creatine Values

Creatine values in males normally range from 0.2 to 0.6 mg/dl; in females, from 0.6 to 1 mg/dl.

Serum Uric Acid Values

Uric acid concentrations in males normally range from 4.3 to 8 mg/dl; in females, from 2.3 to 6 mg/dl.

Diagnostic Studies to Assess the Renal System

A wide variety of tests are available to assess the renal system. They include:
- Urine and blood studies provide information concerning the patient's overall state of health. Urine tests also help evaluate the kidneys' concentrating and diluting ability. Blood tests assess the ability to eliminate waste and maintain homeostasis and help evaluate glomerular and tubular function.
- Clearance tests for filtration, reabsorption, and secretion permit precise evaluation of renal function.
- Imaging tests—radiography, radionuclide imaging, and ultrasonography—visualize renal abnormalities.
- Other specialized tests include cystourethroscopy.

Renal Failure: Assessment Guidelines

Here's a list to help you obtain information for your patient's care plan.
1. Obtain laboratory results:
   - *Serum electrolytes*—check for imbalances, depletion, or excess
   - *Urine electrolytes*—check for wasting of salt or potassium
   - *Urinalysis*—look for tubular casts, protein, red blood cells, and white blood cells
   - *Urine specific gravity*—rule out dehydration and inability to concentrate urine
   - *Urine osmolality, pH, sugar, and acetone*
   - *24-hour urine collection*—check for protein and creatinine clearance
2. Make a physical assessment of the patient:
   - *Check urine* for color, odor, concentration, and sediment
   - *Obtain patient's weight*
   - *Check skin turgor* and mucous membranes
   - *Record any signs* of edema
   - *Check for pain* over kidney to rule out infection
   - *Check for pain*, urgency, and burning upon urination
3. Note any evidence of heart or liver disease as well as any other complications that might affect treatment.
4. Measure all drainage, such as vomitus, diarrhea, draining wounds, fistulas, and GI suction, and send specimen to lab for electrolyte content.

Acute Renal Failure

- Acute renal failure (ARF) affects both kidneys simultaneously. It can result from prerenal, intrarenal, or postrenal causes.
- Acute tubular necrosis (ATN) accounts for about 75% of all cases of ARF. Nephrotoxin-induced ATN usually causes reversible damage, whereas ischemia-induced ATN usually causes irreversible damage.
- The creatinine clearance test provides the most accurate measure of renal function.
- Peritoneal dialysis and hemodialysis are the primary treatments for ARF.
- An accurate patient history and a complete physical examination provide valuable information for formulating nursing diagnosis and implementing an effective care plan.

Causes of Renal Failure

Prerenal: vascular disorders; embolism; infarction; congestive heart failure; shock due to hemorrhage, sepsis, burns, or cardiogenic shock; placenta previa; septic abortion; postpartum hemorrhage; eclampsia; crush injuries; hemolysis; dehydration; surgery in which the aorta or the renal arteries are clamped

Intrarenal: pyelonephritis, nephrosclerosis; acute membranous or membranous glomerulonephritis; Goodpasture's syndrome; systemic lupus erythematosus; periarteritis nodosa; scleroderma; diabetes mellitus; amyloidosis; gout; nephrocalcinosis; hereditary or hypertensive nephropathy; polycystic kidneys; renal hypoplasia; drugs, heavy metals, or industrial solvents; acute tubular or bilateral cortical necrosis; sickle cell disease; radiation nephritis; Wilms' tumor; hypernephroma

Postrenal: congenital collecting duct anomalies; urethral strictures; pyelolithiasis; carcinoma, lymphoma, or Hodgkin's disease; prostatic hypertrophy; urinary tract infection, stones, or obstruction.

Filtering the Flow

In the healthy kidney, the glomerulus, a capillary bed, filters about 120 ml of plasma a minute through its semipermeable membrane. The "glomerular filtrate," or "ultrafiltrate," that results forms the main component of urine. The rate of formation, or glomerular filtration rate (GFR), helps measure the degree of renal function.

These factors decrease the GFR and the formation of urine:
• dehydration or decreased extracellular fluid volume
• hypotension
• decreased cardiac output
• hyponatremia
• renal disease: obstruction and glomerular disease.

These factors increase the GFR and the formation of urine:
• overhydration or increased extracellular fluid volume
• hypertension
• increased cardiac output
• hypernatremia
• renal disease: tubular-medullary disease.

Four Stages of Renal Deterioration

Here's a handy set of definitions to help you assess your patient's renal function.
Diminished renal reserve—Although kidney function as a whole is mildly reduced, the excretory and regulatory functions sufficiently to maintain a normal internal environment.
Renal insufficiency—Some evidence of impaired capacity may appear in the form of mild azotemia, slightly impaired concentrating ability, and anemia. However, these abnormalities are minimal until dehydration, infection, or heart failure put stress on the kidney.
Renal failure—Kidney function has deteriorated to the point of chronic and persistent abnormalities in the internal environment.
Uremic syndrome—Many clinical signs and symptoms may appear in the patient with renal failure.

Dietary Guidelines in Renal Failure

RECOMMENDED NUTRIENTS AND IONS	DURING INITIAL RENAL FAILURE	DURING MAINTENANCE DIALYSIS
High-biologic-value protein	20 to 40 g	In hemodialysis: 1 g/kg body weight (adult), 1.5 g/kg body weight (child) In peritoneal dialysis: Usually unrestricted
Low-biologic-value protein	10 g or less	15 g or less
Carbohydrates	Unrestricted	Unrestricted
Fat	Unrestricted	Unrestricted
Calories	Adult: 45 to 50 kcal/kg Child: 80 kcal/kg	Adult: 45 to 50 kcal/kg Child: 80 kcal/kg
Sodium	Individualized	Individualized (500 to 2,000 mg)
Potassium	Individualized (40 to 70 mEq)	Individualized (40 to 70 mEq)
Calcium	RDA (800 to 1,200 mg)	RDA (800 to 1,200 mg)
Phosphate	Unrestricted in diet (controlled by drugs)	Unrestricted in diet (controlled by drugs)

Dialysis Differences

In patients with renal failure, either hemodialysis or peritoneal dialysis can be used to remove toxic wastes and excess fluid. But each of these kinds of dialysis has its advantages and disadvantages. Here's a comparison.

Peritoneal dialysis:
- Can be done immediately or 48 to 72 hours after the patient has had abdominal surgery
- Doesn't require complex equipment or highly trained personnel but must be done in a hospital
- Doesn't require the administration of heparin or requires it only in small amounts
- May cause more discomfort than hemodialysis
- May cause peritonitis; shock; atelectasis and pneumonia; severe protein loss; perforation of the bowel, bladder, patent urachus, or blood vessel. In addition, the dialysis fluid may be difficult to retrieve from the abdominal cavity.

Hemodialysis:
- Can be done immediately by inserting a catheter into the femoral vein
- Requires complex, expensive machinery and specialized personnel but may be done in a satellite unit, away from the hospital
- Requires good blood vessels for the fistula or shunt
- Takes only 3 to 4 hours
- Requires the administration of heparin, so there's a risk of hemorrhage
- May cause septicemia, loss of an artery (or even a limb), embolism, hepatitis, seizures (from rapid fluid and electrolyte changes), and hypotension.

When Is Dialysis Required?

The patient with renal failure needs some form of dialysis when he develops uremic symptoms, uncontrollable hyperkalemia, hypertension, congestive heart failure, or uncontrollable acidosis.

Restoring Balance

How does peritoneal dialysis work? By a combination of osmosis, diffusion, and filtration, it drains off metabolic waste products and reestablishes fluid and electrolyte balance. Because the membrane is semipermeable, water and the usual solutes can pass back and forth much more freely than large protein and sugar molecules.

With peritoneal dialysis, fluid balance is attained by using solutions of varying tonicity or concentration. A dializing solution with a serumlike electrolyte content, although ordinarily without potassium, is used. If potassium is needed, you can add about 4 mEq/liter. You can also add glucose or dextrose to increase tonicity. This also increases peritoneal clearance, simply because the increased osmolarity (osmotic strength) of the higher concentrate drags more solutes out with the water.

Care in Acute Peritoneal Dialysis

Dialysis will start to solve the problem of azotemia, but it's up to you to see that the strictest sterile technique is used and that the fluid balance and electrolytes swing toward the norm, not away from it.

During the first exchange, take blood pressure and pulse every 15 minutes; take them every hour after that. Take the temperature every 4 hours—and *after* the abdominal catheter is removed. One of your most important jobs throughout is respiratory care. Long periods of hypoventilation enforced by a full abdomen aggravate the risk of pneumonia. Raise the head of the patient's bed, and encourage coughing and deep breathing. Turn the patient from side to side when you can. If he shows marked difficulty in breathing during dialysis, drain the fluid at once and notify the doctor. Also watch for pain, leakage of solution, bleeding, and infection.

What Is Peritoneal Dialysis?

If the kidneys continue to fail despite other measures, dialysis may be necessary. Today, with disposable catheters and commercially manufactured dialyzing solutions, peritoneal dialysis has become a simple and easily accessible method for removing toxic wastes and excess fluid. It's often used to prevent uremia during diagnostic evaluation; while getting the patient in shape for surgery; and in chronic renal failure, especially in patients who lack vascular access or who are hemodynamically unstable. This form of dialysis infuses a solution into the peritoneal cavity through a closed drainage system. It uses the peritoneum as a dialyzing membrane to replace the malfunctioning kidneys. The peritoneum's filtering surface—about 22,000 square centimeters—approximates the surface of the glomerular capillaries. This strong, smooth, colorless, serous membrane lines the walls of the abdominal cavity with its own parietal surface and wraps the abdominal organs inside its visceral one. The peritoneum is continuous in the male, but in the female the ovaries and the fallopian tube jut through it. In peritoneal dialysis the fluid is instilled into the peritoneal cavity between these two layers and is left for a controlled length of time.

Diet and dialysis
Diet planning for the dialysis patient must achieve several goals. The diet must:
- provide adequate caloric intake to achieve ideal body weight
- provide adequate protein to maintain positive nitrogen balance (for example, use of high-biologic-value protein and essential amino acids)
- control serum sodium to prevent hypotension, hypertension, or excessive thirst
- control serum potassium to prevent hyperkalemia or hypokalemia

Continued

What Is Peritoneal Dialysis?
Continued

- control serum phosphorus to augment phosphate-binding drugs
- control fluids to prevent hypertension or edema
- provide vitamin and mineral supplements, as needed.

The diet must also be practical, palatable, and acceptable to the patient

Use of low-biologic-value proteins provides few essential amino acids and furnishes more urea nitrogen than high-biologic-value proteins during metabolism, which increases the BUN level.

Limited Access
Patients on hemodialysis must have an access route to deliver their blood to the dialysis machine. Here are several ways to provide such access:

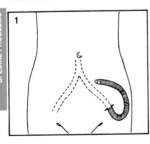

1. *Femoral vein catheterization* provides quick, temporary (about 1 week) access to the circulation in emergencies. But this access site increases the risk of infection or of femoral artery puncture. When the catheter's in place, the patient must remain supine in bed to avoid dislodging the catheter, damaging the vein, or obstructing the blood flow.

Continued

Limited Access
Continued

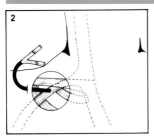

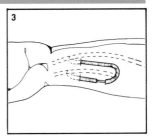

2. *Subclavian vein catheterization* also provides quick, temporary (1 to 2 weeks) access to the circulation in emergencies but with less risk of infection than femoral vein catheterization. However, it is contraindicated in pulmonary hypertension and carries a risk of pneumothorax on insertion. The patient may ambulate with this type of catheter.

3. *Arteriovenous (AV) shunt* provides access to arterial and venous blood vessels without repeated catheterization or venipuncture. It can be used temporarily or for longer periods (average 7 to 10 months) but restricts the patient's activity in the affected arm or leg. This access site increases the patient's risk of infection and clotting and may accidentally separate, causing severe hemorrhage and possibly death.

Have the patient keep the affected arm or leg straight and elevated for 2 to 3 hours after insertion. Also, teach the patient to clean the shunt daily. Carefully examine the skin around the shunt for redness, tenderness, drainage, and erosion over the tubing. The tubing should feel warm, and you should feel a bruit or thrill. If the tubing is cool, if there's no bruit or thrill, or if you can see blood separation in the tubing, the shunt may be clotted. Report this to the doctor immediately.

Continued

Limited Access
Continued

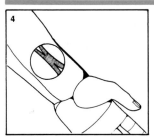 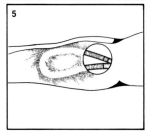

4. *AV fistula* is the preferred access route for patients on hemodialysis because it's less likely to be damaged or disconnected by trauma and because it carries a lower risk of infection than the AV shunt. This access site can be used for long periods (3 to 4 years) but is contraindicated in patients with small veins.

One common problem with a fistula is steal syndrome, numbness, tingling, and coldness in the access arm. Fistulas may also close up, may lead to aneurysm formation, or may bleed.

Like a shunt, a fistula should be inspected daily for signs of infection and presence of a bruit or thrill; and the fistula arm shouldn't be used for taking blood pressure, for drawing blood, or for intravenous therapy. Patients with a fistula should know how to apply pressure until bleeding stops.

5. *AV vein graft* provides a long-term (average 2 years) access route that allows unrestricted use of the grafted arm or leg. Like the AV fistula, the graft increases the patient's risk of steal syndrome (possible numbness, tingling, and coldness below the fistula site from arterial insufficiency in atherosclerotic or diabetic patients). It also increases the risk of clotting from hypotension and of infection in tissues surrounding the graft.

Three Forces in Dialysis

Osmosis: Fluid crossing a membrane moves from a dilute solution to a more concentrated one (as fresh water is drawn into brine). Peritoneal dialysis promotes fluid balance by using a dialyzing solution with a serumlike electrolytic content (except that it ordinarily contains no potassium, because failing kidneys can't excrete it). So you usually add the appropriate amount of potassium with glucose or dextrose to increase tonicity, according to the patient's fluid status. Also, because peritoneal capillaries produce fibrin when irritated, you may need to add heparin to prevent fibrin formation and plugging of the catheter.

Diffusion: Particles crossing a membrane move from a more concentrated one. In peritoneal dialysis, the accumulated waste particles in the blood freely diffuse into the dialyzing bath. Glucose can draw water (and solutes) out of the body, but its molecules are too big to get in readily. So the dialyzing fluid withdraws only small waste particles and water—the amount drawn depends on the amount of glucose put in.

The patient may also need a replacement for ascorbic acid and folic acid lost in dialysis. Once the solution has been infused, it's usually left in the peritoneal cavity for about 20 to 25 minutes. Longer dwell time could result in fluid overload.

Filtration: Hydrostatic pressure pushes the body fluid out through the body's membranes, then into the abdomen and through the tube, into the collecting bottle below. Hydrostatic pressure depends on the fact that water must seek its own level.

Complications of Dialysis

Blood loss
Hemorrhage
Hemolysis
Pyrogenic reaction
Hypotension or hypertension
Dysrhythmias

Muscle cramps
Disturbed sensorium
Anemia, possibly with splenomegaly
Hepatitis
Accelerated atherogenesis

Managing Hemodialysis Complications

COMPLICATIONS/POSSIBLE CAUSES	NURSING CONSIDERATIONS
Internal hemorrhage/ Excessive heparinization	Reduce the heparin dose, or use minimal or regional heparinization. Observe the patient for signs of internal bleeding, increased respirations, and decreased temperature. If ordered, perform a blood transfusion.
External hemorrhage/ Line disconnection	Observe blood lines for leakage. Keep clamps and blood pressure cuff nearby.
Dialysis disequilibrium syndrome/Rapid shift of fluid and electrolyte levels	Reduce blood flow, and inform the doctor immediately. He may order diazepam or phenytoin sodium, or he may discontinue dialysis.
Hypotension/Septic shock, decreased cardiac output, or reduced blood volume from extracorporeal circulation	Infuse normal saline solution, as necessary, to restore blood volume. If ordered, administer mannitol, plasma, or albumin. Monitor blood pressure.
Dysrhythmia or angina/ Rapid shift of fluid and electrolyte levels, reduced blood volume from extracorporeal circulation, or reduced hematocrit level	As ordered, give sodium polystyrene sulfonate (Kayexalate) for hyperkalemia, blood transfusions for decreased blood volume, or antiarrhythmics.

What Happens in Hypovolemic Shock

Vascular fluid volume loss causes the extreme tissue hypoperfusion that characterizes hypovolemic shock. *External fluid loss* results from severe bleeding or from severe diarrhea, diuresis, or vomiting. Causes of *internal fluid loss* include internal hemorrhage (such as GI bleeding) and third-space fluid shifting (as in diabetic ketoacidosis). Inadequate vascular volume leads to decreased venous return and cardiac output. The resulting drop in arterial blood pressure activates the body's compensatory mechanisms in an attempt to increase vascular volume. If compensation's unsuccessful, decompensation and death may rapidly ensue.

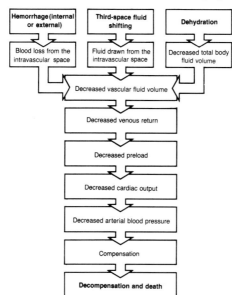

Recognizing Internal Fluid Loss

LESS THAN 15% OF CIRCULATING FLUID VOLUME (750 ML)

Possible causes
- Fractures
- Bleeding ulcer

Signs and symptoms
- Thirst
- Slight increase in pulse rate
- Slight increase in respiratory rate
- Normal blood pressure
- Possible increased nail bed blanching time (suggesting vasoconstriction)
- Negative tilt test
- Decreased urine output

20% to 25% OF CIRCULATING FLUID VOLUME (1,000 TO 1,250 ML)

Possible causes
- Pelvic or long-bone fractures
- Liver or spleen trauma

Signs and symptoms
- Tachycardia (more than 120 beats/minute)
- Tachypnea (more than 30 respirations/minute)
- Hypotension
- Narrowing pulse pressure (systolic pressure decreases more rapidly than diastolic pressure)
- Decreased nail bed blanching time
- Positive tilt test

30% to 40% OF CIRCULATING FLUID VOLUME (1,500 TO 1,800 ML)

Possible causes
- Multiple trauma with fractures
- Spleen rupture
- Thoracic injuries
- Vascular injuries

Signs and symptoms
- Extremely low urine output
- Cool, clammy skin
- Tachycardia
- Hypotension
- Narrowing pulse pressure
- Anxiety
- Thirst
- Decreased level of consciousness

40% to 50% OF CIRCULATING FLUID VOLUME (2,000 TO 2,300 ML)

Possible causes
- Laceration of major artery or vein
- Multiple trauma
- Severe liver, spleen, or kidney injury

Signs and symptoms
In addition to the signs and symptoms previously listed, expect to find:
- Weak, thready pulse; may not be palpable
- Dyspnea
- Coma

How Bowel Obstructions Lead to Shock

How can a bowel obstruction cause hypovolemic shock? Consider this sequence of events:

- Because the bowel is blocked, gas, fluid, and other intestinal contents accumulate.
- Feeding on these substances, normal bacteria proliferate. Their activity produces more gas, which promotes further bowel distention.
- Transient, but violent, peristalsis may further elevate bowel pressure.
- As the bowel distends, it secretes more fluids, removing fluids and electrolytes from the intravascular compartment and depositing them in the bowel lumen. This type of third-space fluid shift causes hypovolemia and may lead to hypovolemic shock. These secretions also contribute to increased bowel pressure, which hampers the bowel's ability to reabsorb fluid.
- Pressure from bowel contents compresses intestinal wall tissue, impairing circulation and weakening the tissue. Consequently, fluids and electrolytes leak into the intestinal walls, and edema increases.
- When bowel pressure exceeds venous and arteriolar pressures, lymphatic and capillary statis occurs and drainage is impaired.
- Decreased blood flow to the bowel leads to necrosis.

As you can see, intestinal obstruction causes decreased blood volume and hemoconcentration from loss of intravascular fluid. As with all types of shock, the decreased circulating fluid volume may lead to renal insufficiency and diminished urine output. Unless relieved, a bowel obstruction is fatal.

Hypovolemia (Not From Hemorrhage)

(Decreased circulating fluid volume due to loss of water and electrolytes)

CAUSES	SIGNS AND SYMPTOMS	LABORATORY FINDINGS
Decreased water intake Fluid loss due to diarrhea, fever, vomiting Systemic infection Impaired renal concentrating ability Fistulous drainage Severe burns Hidden fluid in body cavities	Increased... —pulse rate —respiratory rate Decreased... —blood pressure —body weight Weak and thready peripheral pulses Thick, slurred speech Thirst Oliguria (diminished urine output compared with fluid intake) Anuria Dry skin	Increased... —red blood cell count —hemoglobin concentration —packed cell volume —serum sodium concentration —urine specific gravity

Assessing Fluid Replacement Therapy

How's your shock patient responding to fluid replacement? Use this checklist as an assessment guide. If you can answer "yes" to most or all of these questions, your patient's probably on the way to recovery.
Remember: Normal values vary among patients. Always compare your readings with your patient's baseline measurements.

• Is your patient's urine output increasing? (Expect urine output to measure more than 30 ml/hour as his condition improves.)

• Is his urine specific gravity measurement decreasing?

• Is his pulse rate approaching normal?

• Is his blood pressure increasing and stabilizing?

• Does he appear alert? Does he feel less anxious?

• Are his arterial blood gas (ABG) measurements improving? As your patient improves, expect his ABG and pH values to approach normal.

• Is his central venous pressure (CVP) rising? (Expect CVP to be between 5 and 15 cmH$_2$O.)

• Is his skin color normal, rather than pale or gray? Is skin turgor normal?

Two Pumps in One

Consider the heart as two pumps, one on the right and one on the left. Each side depends on the other for its own efficiency since the failure of one pump will eventually affect the other. If you understand the symptoms associated with right- and left-sided heart failure, as shown here, you will sharpen your assessment skills and be more helpful to your patient.

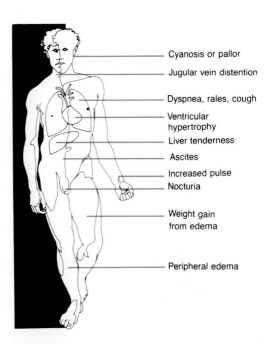

- Cyanosis or pallor
- Jugular vein distention
- Dyspnea, rales, cough
- Ventricular hypertrophy
- Liver tenderness
- Ascites
- Increased pulse
- Nocturia
- Weight gain from edema
- Peripheral edema

Checking the Hepatojugular Reflux

In patients with right-sided heart failure, you can detect venous congestion early by evaluating the hepatojugular reflux. Place the patient in a semi-Fowler's position so that she is elevated enough for the blood column of her jugular vein to be visible above the clavicle.

Have the patient relax and breathe normally. Now place your hand on her abdomen. As you press firmly toward the right upper quadrant and under the costal margin, watch the blood column of her jugular vein. If she has venous congestion, the pressure you apply will displace a small amount of blood from the liver and cause her jugular vein to distend about 1 to 2 cm.

How Electrolyte Imbalances Affect Heart Function

Sodium, potassium, and calcium are the three electrolytes that most strongly influence cardiac conduction and contractility. Here's how imbalances of these electrolytes affect your patient's heart function.
Hyperkalemia. When serum potassium levels are high, the heart becomes flaccid, or sluggish. Heart rate slows and contractions become less forceful. A large potassium excess causes ventricular fibrillation or cardiac arrest.
Hypokalemia. If potassium levels drop below normal, contractions become more forceful but erratic, resulting in supraventricular and ventricular ectopic beats. Like hyperkalemia, hypokalemia can cause ventricular fibrillation.
Hypercalcemia. A small calcium excess increases contractility. A large excess may cause erratic contractions, heart block, or cardiac arrest.
Hypocalcemia. Like hyperkalemia, a calcium deficit reduces contractility and slows heart rate. A large deficit may cause heart block, premature beats, or ventricular fibrillation.
Hypernatremia. Too much sodium causes depressed cardiac function. For unknown reasons, an excess of sodium ions reduces calcium's normal ability to stimulate forceful myocardial contractions.

How CHF Produces Fluid Imbalance

Congestive heart failure produces fluid imbalance via several interlocking mechanisms. For example, edema develops because:
- decreased renal blood flow increases secretion of aldosterone (as much as triple the normal amount), causing sodium and water retention
- liver congestion, a result of venous blood congestion, may aggravate edema by preventing inactivation of aldosterone and ADH (which are normally inactivated by the liver)
- excessive venous blood pooling increases hydrostatic pressure and so shifts fluid from the intravascular to the interstitial compartment.

At the same time, electrolyte imbalances develop because:
- excessive aldosterone secretion promotes potassium excretion and leads to potassium deficit
- diuretics also waste potassium and lead to its deficit
- mercurial diuretics, although rarely used, cause selective excretion of more chloride ions than sodium ions; chloride loss brings compensatory increase in bicarbonate ions, hence metabolic alkalosis
- pulmonary congestion interferes with the elimination of carbon dioxide from the lungs and leads to respiratory acidosis
- extensive use of diuretics plus severely restricted sodium intake may lead to sodium deficit.

Balancing Act

The electrolytes sodium and potassium play a crucial role in neuromuscular functioning through the processes of depolarization and repolarization. Without depolarization-repolarization, the electrical impulse initiated by the SA node can't travel throughout the conduction system and produce a rhythmic, healthy heartbeat.

In the resting cell (Figure 1), anions accumulate along the inner surface of the cell membrane as cations accumulate along the outer. Stimuli—such as heat, cold, electricity, mechanical damage, or any other factor that temporarily disrupts the normal resting state—cause the cell membrane to become very permeable to sodium. Sodium rushes into the cell, reversing the original resting potential, and potassium moves out. This shift is called depolarization (Figure 2).

Within a few milliseconds, sodium moves out of the cell again and potassium returns to its cellular compartment. This shift is called repolarization (Figure 3). Repolarization returns the cell to its electrical resting potential.

This electrolyte exchange is transmitted along the axon, causing nerve conduction and muscle contraction. An electrolyte imbalance will affect the synchronization of neuromuscular function, which can lead to irritability of muscles, nerves, and the heart. If the nerve stimulus is transmitted during the refractory period of the muscle, there'll be no contraction. If the contraction is sustained, tetany may result.

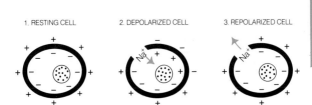

1. RESTING CELL   2. DEPOLARIZED CELL   3. REPOLARIZED CELL

Treatment Goals of Congestive Heart Failure

Treatment of congestive heart failure aims to:
1. eliminate the cause whenever possible (as in surgical correction of incompetent or stenosed valves)
2. establish cardiac output adequate for metabolic needs by:
—increasing the force of myocardial contraction with digitalis and
—reducing the demand on cardiac output by reducing physical activity
3. correct fluid congestion and electrolyte imbalances by reducing circulatory blood volume with diuretics; by reducing sodium concentration with diuretics and dietary restriction; and by maintaining normal serum potassium levels with potassium supplements and potassium-sparing diuretics in combination with thiazides.

After CPR

When sodium bicarbonate (usually a 7.5% concentration) is administered in large quantities during or immediately after cardiopulmonary resuscitation (CPR), your patient may be a candidate for metabolic alkalosis. Astute nursing care during and immediately after CPR can prevent or detect this serious complication. Follow these guidelines:
• Carefully monitor blood gases. Suspect metabolic alkalosis when you see pH above 7.45, HCO_3^- above 26 mEq/liter, and $Paco_2$ above 45 mm Hg (if compensating).
• After the code, note your patient's rate, depth, and pattern of breathing. Slow, shallow, and possibly irregular breathing may indicate metabolic alkalosis.
• Watch for other signs of metabolic alkalosis, such as restlessness, irritability, confusion, hypertonic muscles, twitching, and tetany. In severe cases, seizures and coma may result.

Questions to Ask Your Patient Before Surgery

Here are some important questions to answer about your patient before he undergoes surgery.
- What are the patient's preexisting health problems?

- Does he have a cardiovascular problem?

- Was edema present on admission?

- Did the patient have problems that might lead to dehydration, such as vomiting, diarrhea, anorexia, diaphoresis, or hyperventilation?

- How old is the patient? The very old and the very young are exceedingly vulnerable to complications and need close assessment.

- What is the patient's fluid output? If it's less than 500 ml/day, he may be dehydrated or may have renal insufficiency.

- Were laboratory tests ordered? If so, monitor serum electrolyte, hemoglobin, hematocrit, blood urea nitrogen, and creatinine levels.

Special Consideration

After Surgery: Watch for fluid and electrolyte imbalance by checking laboratory findings and watching for signs of dehydration. When assessing the patient, consider the type of surgery he's had. For instance, the patient who's had gastrointestinal surgery is apt to lose electrolytes in amounts sufficient to cause imbalance.

Fluids Critical in Pediatric Surgery

Surgery risks increase in babies and children depleted of water, sodium, and potassium or in those with acid-base disorders.
So *before surgery:*
- Assess and correct any fluid imbalance.
- Take vital signs for baseline.
- Determine electrolyte levels.
- Rapid breathing and perspiration intensify insensible water loss. Use a mist tent to minimize it. Some authorities recommend increasing fluid replacement by 10% for each 1° F. rise in temperature.
- If urinary output or kidney function is poor, establish flow with a hypotonic or isotonic infusion. With diarrhea or vomiting, determine the fluid and electrolyte replacement required. If the baby is dehydrated, start with a hypotonic saline solution, then use an isotonic dextrose 5% in water (D_5W) solution followed by a ⅓ normal saline solution. Use a hypotonic solution for infants since their kidneys can't concentrate urine well. After you establish urinary flow, use a multiple electrolyte solution.
- If a child is to receive nothing by mouth before surgery, don't withhold liquids longer than 4 to 6 hours. For the hours remaining before surgery, give I.V. infusions. Then watch for overhydration.

During surgery:
Use glucose solutions to replace fluid loss. Don't use multiple electrolytes unless the patient's blood values and symptoms show a clear need.

After surgery:
Immediately withhold multiple electrolytes, especially in older children, in favor of glucose or glucose with saline solution. Stress and trauma at this time sharply reduce water output. Postoperative hemorrhage, sweating, or hyperventilation can produce or aggravate dehydration. If gastric suction is used, replace the fluid it removes volume for volume. Use D_5W in a ½ normal saline solution, with 1 mEq/liter of potassium added for every 100 ml of gastric juice removed. Know how to treat the three kinds of dehydration:
- Isotonic or simple (proportional water and salt loss). This is the most common kind and the easiest to treat with rapid fluid replacement. Calculate the loss over 24 hours and, in severe cases, replace half in the first 8 hours and half over the next 16 hours.
- Hypertonic (hypernatremia: more water lost than salt). This results from water loss through liquid stools, low water or high solute intake, rapid breathing, poor renal function, or any combination of these.

Continued

Fluids Critical in Pediatric Surgery
Continued

Increase fluid volume, but watch for water intoxication and seizures from serum sodium levels lower than 100 mEq/liter and low calcium levels. *Don't attempt rapid full correction!* Gradually reduce the excess sodium over at least 24 hours. Use D_5W or a physiological salt diluted by ¼ or ⅓. Calcium gluconate may be added to prevent tetany, twitching, seizures, and cardiac dysrhythmias. Later, when urinary flow returns to normal, about 3 mEq/kg of potassium may be added.

- Hypotonic (hyponatremia: more salt lost than water). This imbalance can result from copious water intake, electrolyte-free fluid infusions, or excessive sweating. Low sodium levels impair cardiac and renal functions and may produce cerebral edema with seizures.

Again, *don't attempt rapid full correction.* Restrict fluid intake or give saline solution slowly and carefully. Too-rapid correction can cause congestive heart failure and pulmonary edema, especially in a child with cardiac problems.

Special Consideration

Assessing elderly patients challenges your skill. Learn to distinguish between the normal signs of aging and the more severe signs of fluid imbalance. Watch for loose skin, dry, cracked lips, sunken eyes, and hollow cheeks. Don't rely on generalizations. Know what is normal for your patient.

The chief vulnerabilities of the aged are fluid imbalance, electrolyte imbalance, and acidosis or alkalosis. When the body is no longer sturdy, the respiratory, renal, cardiac, and gastrointestinal systems undergo physiologic changes that affect function and fluid balance.

Some Commonly Used I.V. Products

Before starting an I.V., always make sure you check the doctor's orders against the I.V. bottle from the pharmacy. Does it contain the correct fluid? Is the amount correct? Do you have all the additives ordered by the doctor? If additives are used, note and label their names, dose, rate of administration, and expiration date and time on the I.V. bottle.

PRODUCT	ELECTROLYTES	
	ELEMENT SUPPLIED	AMOUNT SUPPLIED
Dextrose solutions		
2½%		
5%		
Saline solutions		
0.45%	Sodium	77 mEq/liter
	Chloride	77 mEq/liter
0.9%	Sodium	154 mEq/liter
	Chloride	154 mEq/liter
3%	Sodium	513 mEq/liter
	Chloride	513 mEq/liter
5%	Sodium	855 mEq/liter
	Chloride	855 mEq/liter

Nursing Tip

When you have several patients receiving I.V. fluids, a quick way to check on the absorption rate is to attach a piece of adhesive tape lengthwise to each bottle. At the top of the tape, mark the time the solution was hung. At the bottom of the tape, mark the time the solution should be absorbed. Midway between these two labels, mark the time when half the amount of the solution should be absorbed. With these markings, you'll be able to see at a glance whether the solutions are being absorbed on schedule.

—Sylvia E. Platt, RN

If the patient will be receiving several different I.V. fluids during the day, also check the solution bottle to make sure you're giving the right fluid in the right sequence.

INDICATIONS	PRECAUTIONS AND NURSING IMPLICATIONS
Maintains water balance and corrects imbalance; supplies calories as carbohydrates	• Electrolyte-free solutions may cause peripheral circulatory collapse and anuria in patients with sodium deficiency and may aggravate hypokalemia. Do not administer with blood.
Fluid replacement, dehydration, sodium depletion; low salt syndrome (hyponatremia)	• Use sodium solutions with caution in edematous patients and those with heart, renal, or hepatic disease. Administer slowly.

Continued

Nursing Tip

When the I.V. runs dry and you need to hang another bottle but air is in the tubing, here's a quick way to change bottles without breaking a closed system or introducing a needle into it. First, slide the drop regulator down the tubing and close it. Remove the empty bottle and hang the new one. Squeeze the drip chamber to get fluid into it. Then grasp the tubing just above the drop regulator, using your thumb and a hard object to compress it. Move your hand up the tubing. Fluid will push air into the drip chamber. Then release your grasp and turn on the I.V.

—P. Seibel, RN

Some Commonly Used I.V. Products
Continued

PRODUCT	ELECTROLYTES	
	ELEMENT SUPPLIED	AMOUNT SUPPLIED
Dextrose with saline		
5% dextrose and 0.45% NaCl	Sodium	77 mEq/liter
	Chloride	77 mEq/liter
5% dextrose and 0.9% NaCl	Sodium	154 mEq/liter
	Chloride	154 mEq/liter
Dextrose 2½%, 5%, or 10% in Ringer's lactate	Sodium	130 mEq/liter
	Potassium	4 mEq/liter
	Calcium	3 mEq/liter
	Chloride	109 mEq/liter
	Lactate	28 mEq/liter
Ringer's (plain)	Sodium	147 mEq/liter
	Potassium	4 mEq/liter
	Calcium	4.5 mEq/liter
	Chloride	155.5 mEq/liter
Ringer's lactate U.S.P. (plain) (Hartmann's solution)	Sodium	130 mEq/liter
	Potassium	4 mEq/liter
	Calcium	3 mEq/liter
	Chloride	109 mEq/liter
	Lactate	28 mEq/liter

Nursing Tip

To remove a stubborn needle from I.V. tubing or a syringe, wrap one end of some Penrose drain tubing around the I.V. needle's hub and the other end around its adapter. As you twist in opposite directions, the Penrose tubing "grabs" the line to help you pull out the needle.

—Nancy Perez Diatima, RN

INDICATIONS	PRECAUTIONS AND NURSING IMPLICATIONS
Fluid replacement, caloric feeding, dehydration, sodium depletion Replacement of surgical or GI loss, dehydration, sodium depletion, acidosis, diarrhea, burns	
	• Check urine flow before giving potassium.
Dehydration, sodium depletion, replacement of GI loss	• Check urine flow before giving potassium.
Replacement of surgical and GI loss, dehydration, sodium depletion, acidosis, diarrhea, burns	• Check urine flow before giving potassium.

Continued

Nursing Tip

If you suspect I.V. infiltration in a patient with difficult veins, turn on a flashlight and hold it against his skin, directly over the suspicious site.

If I.V. fluid has infiltrated the tissue, the beam will highlight the size of the infiltration. If no fluid has infiltrated, only a small halo will appear around the flashlight.

Using this trick can save you from having to do extra checks. Then, if necessary, you can stop the I.V. before the infiltration gets worse.

—Betty Woodfin, RN

Some Commonly Used I.V. Products
Continued

PRODUCT	ELECTROLYTES	
	ELEMENT SUPPLIED	AMOUNT SUPPLIED
M/6 sodium lactate injection U.S.P. (plain)	Sodium Lactate	167 mEq/liter 167 mEq/liter
5%, 10%, 20% mannitol in 0.45% NaCl	Sodium Chloride	77 mEq/liter 77 mEq/liter
Dextran 40 10% injection with 5% dextrose		
10% injection with 0.9% NaCl	Sodium Chloride	154 mEq/liter 154 mEq/liter
Dextran 70 6% injection with 5% dextrose		
6% injection with 0.9% NaCl	Sodium Chloride	154 mEq/liter 154 mEq/liter

SURGERY

INDICATIONS	PRECAUTIONS AND NURSING IMPLICATIONS
Severe metabolic acidosis (raises bicarbonate levels)	
Test for renal function (for example, oliguria due to tubular necrosis); diuretic therapy for intoxications, edema, and ascites	• Do not give to patients with impaired renal function who fail to respond to the test dose or to those with severe congestive heart failure, metabolic edema, or head injuries. Low room temperature may cause crystallization. Use blood filter set to prevent infusion of mannitol crystals.
Provides plasma volume expansion and early fluid replacement in shock when whole blood or blood products are not available or when urgency does not allow time for cross matching	• Watch for allergic reaction, such as mild urticaria. Stop infusion at first sign of reaction.
Provides plasma volume expansion and early fluid replacement in shock and hypovolemia when whole blood or blood products are not available or when urgency does not allow time for cross matching	• Monitor urine flow: if oliguria or anuria occurs, stop infusion. Contraindicated in dehydration or kidney disease. Do not exceed 2 g/kg of body weight per 24 hours.

Nursing Tip

When an I.V. is to be discontinued after the patient has absorbed all the fluid, tape a Band-Aid and a packaged alcohol swab on the I.V. bottle. Not only will this save time, but it confirms at a glance the discontinue order.
—Patricia Wilson, RN

A Vital Loss

Burn damage increases capillary permeability. This increase and the inflammatory process cause leakage into the interstitial space. Since water, electrolytes, and albumin molecules are small, a greater number of them than of blood cells (white blood cells and platelets) or of large protein molecules like globulins, which remain in the vessels, are lost from the vascular space.

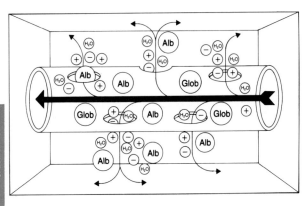

Where Edema Forms

Deep partial- and full-thickness burns destroy all function in the epidermis. Beneath that layer, fluid escapes from dilated capillaries into damaged tissue, causing edema. These drawings show a cross section of normal skin, skin at the time of injury, and skin 24 hours after the burn.

Burned skin may not expand with edema. Then, edema compresses the underlying vessels. For example, such edema of the neck and chest endangers the airway and respiratory function. Decompression escharotomy is then indicated.

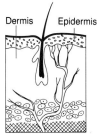
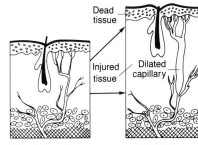

Guidelines for Fluid Replacement

While fluids are being replaced, assess your patient's response and use it to titrate the rate of fluid replacement. But use caution in giving massive fluid replacement to elderly and pediatric patients and to anyone with a history of heart failure. Osmotic diuretics or low-dose dopamine infusion may be necessary to maintain adequate urine output and to prevent fluid overload in these patients, or to aid myoglobin clearance in patients with deep-muscle damage.

ASSESSMENT FACTORS	NURSING CONSIDERATIONS
Intake and output (hourly)	Maintain minimum urine output at 30 to 50 ml/hr in an adult, 0.5 to 1 ml/kg/hr in a child, or 70 to 100 ml/hr in a patient with a deep burn injury affecting muscle tissue, to prevent renal failure from myoglobinuria.
Vital signs (every 15 minutes to hourly)	Maintain the patient's blood pressure above 90/60. Increase the fluid infusion rate and notify the doctor if the patient's blood pressure drops more than 20 mm Hg below baseline or if his pulse rises above 110 beats/minute.
Mental status (continuously)	Note changes such as restlessness, confusion, or agitation in a previously quiet patient, indicating poor cerebral perfusion.
Body weight (same time daily)	Expect the patient to gain weight during the first 48 to 72 hours because of third-space fluid shifting. Thereafter, expect the patient's weight to slowly decrease toward normal dry weight.

Continued

Guidelines for Fluid Replacement
Continued

ASSESSMENT FACTORS	NURSING CONSIDERATIONS
Respiratory status (hourly)	Check the patient's breath sounds for rales and note dyspnea, which may indicate fluid overload. If rales or dyspnea is present, decrease the fluid administration rate and notify the doctor.
Cardiac status (frequently)	Monitor the EKG continuously with an elderly patient or a patient with a history of heart failure, and auscultate heart sounds at least hourly; notify the doctor if rales, dysrhythmias, or abnormal heart sounds—which may indicate fluid overload—develop.
Blood tests (at least daily) • **hematocrit** • **sodium**	Notify the doctor if the patient's values are elevated (possibly indicating underhydration) or decreased (possibly indicating overhydration). He may change the infusion rate and the type of fluid administered and may order a blood transfusion. (Note: Expect an initial rise in hematocrit values.)
Urine specific gravity (every 4 hours)	If elevated, expect to increase the infusion rate; if decreased, to decrease the infusion rate.

Guidelines for Fluid Replacement

FORMULA	ELECTROLYTE-CONTAINING SOLUTION	COLLOIDS	DEXTROSE
FIRST 24 HOURS POSTBURN			
Baxter (Parkland)	Ringer's lactate—4 ml/kg/% burn	Not used	Not used
Hypertonic sodium solution	Volume of fluid containing 250 mEq of sodium per liter to maintain hourly urinary output of 30 ml	Not used	Not used
Modified Brooke	Ringer's lactate—2 ml/kg/% burn	Not used	Not used
Burn budget of F.D. Moore	Ringer's lactate—1,000 to 4,000 ml; 0.5 normal saline—1,200 ml	7.5% of body weight	1,500 to 5,000 ml
Evans	Normal saline—1 ml/kg/%burn	1 ml/kg/% burn	2,000 ml
Brooke	Ringer's lactate—1.5 ml/kg/% burn	0.5 ml/kg/% burn	2,000 ml

Continued

Guidelines for Fluid Replacement
Continued

FORMULA	ELECTROLYTE-CONTAINING SOLUTION	COLLOIDS	DEXTROSE
SECOND 24 HOURS POSTBURN			
Burn budget of F.D. Moore	Ringer's lactate—1,000 to 4,000 ml; 0.5 normal saline—1,200 ml	2.5% of body weight	1,500 to 5,000 ml
Evans	½ of first 24-hour requirement	½ of first 24-hour requirement	2,000 ml
Brooke	½ to ¾ of first 24-hour requirement	½ to ¾ of first 24-hour requirement	2,000 ml
Parkland	Not used	20% to 60% of calculated plasma volume (within 24 to 32 hours)	As necessary to maintain urinary output
Hypertonic sodium solution	⅓ isotonic salt solution orally, up to 3,500-ml limit	Not used	Not used
Modified Brooke	Not used	0.3 to 0.5 ml/kg/% burn	As necessary to maintain urinary output

Lab Values in Pancreatitis

Pancreatitis, an acute inflammation of the pancreas, is often associated with alcoholism and biliary tract disease. The fundamental mechanism in pancreatitis appears to be autodigestion of pancreatic tissue and blood vessels by pancreatic enzymes. Pancreatic enzymes destroy tissues and blood vessels, causing fat necrosis and liquefaction. The result—edematous or hemorrhagic pancreatitis. Edematous pancreatitis is generally self-limiting and subsides within 2 to 3 days. However, hemorrhagic pancreatitis characteristically produces hyperglycemia, hypocalcemia, persistent ileus, and accumulation of necrotic debris in and around the pancreas. Hemorrhagic pancreatitis is a medical crisis that is fatal in 50% of patients.

Serum amylase: elevated—greater than 500 to 2,000 Somogyi units/dl; rises in 48 to 72 hours and decreases by the third day of the attack. Draw blood sample for this test *before* all other diagnostic or therapeutic measures.
Urine amylase: elevated—more than 80 amylase units per hour
Serum bilirubin: elevated—more than 0.2 to 1 mg/dl
Blood urea nitrogen: elevated—more than 5 to 25 mg/dl
Serum calcium: decreased—less than 8 mg/dl or less than 4.5 to 5.5 mEq/liter
Creatinine: elevated—more than 0.5 to 1.5 mg/dl

Serum lipase: elevated—more than 0.1 to 1.5 units/ml
Serum lipids: hyperlipidemia—more than 400 to 800 mg/dl
Serum potassium: elevated—more than 3.5 to 5 mEq/liter in excess pancreatic tissue destruction, or decreased—less than 3.5 to 5 mEq/liter in nasogastric drainage or vomiting
Blood glucose: elevated—more than 90 to 120 mg/dl (Folin-Wu)
Serum triglycerides: elevated—more than 25 to 150 mg/dl (Folin-Wu)
White blood cells (leukocytes): elevated—12,000 to 20,000
(neutrophils): elevated—more than 50% to 75%.

Understanding IVH Solutions

After determining your patient's nutritional deficiencies, the doctor prescribes IVH to best meet your patient's needs. To understand the functions of each component, review this chart.

IVH COMPONENT	PURPOSE
50% dextrose in water	Provides calories needed for metabolism
Amino acids	Supply protein needed for tissue repair
Potassium	Functions in cellular activity and tissue synthesis
Folic acid	Functions in deoxyribonucleic acid (DNA) formation
Vitamin D	Maintains serum calcium levels and functions in bone metabolism
Vitamin B complex	Aids final absorption of carbohydrates and protein
Vitamin K	Helps prevent bleeding disorders
Vitamin C	Promotes wound healing
Sodium	Helps control water distribution to maintain fluid balance
Chloride	Helps regulate acid-base equilibrium and maintain osmotic pressure
Calcium	Promotes blood clotting; aids teeth and bone development

Continued

Understanding IVH Solutions
Continued

IVH COMPONENT	PURPOSE
Phosphate	Minimizes threat of peripheral paresthesia
Magnesium	Helps absorb carbohydrates and protein
Acetate	Prevents metabolic acidosis
Trace elements (zinc, cobalt, manganese)	Promote wound healing and red blood cell synthesis

Special Consideration

To enhance the success of IVH therapy and reduce its risks, observe the following guidelines:
- Before beginning therapy, make sure the doctor's confirmed correct catheter placement with a chest X-ray.
- Use an infusion pump to administer IVH solution. Doing so helps maintain an accurate flow rate and minimizes the risk of clotting in the catheter.
- Maintain meticulous fluid intake and output records.
- Weigh your patient daily. He may gain up to ¾ lb (0.35 kg) through protein synthesis; attribute a greater weight gain to the fluids he's receiving.
- Measure his urine glucose, ketones, and specific gravity every 4 to 6 hours.
- Monitor his vital signs at least once every 4 hours or more often for patients with kidney or liver disease.
- Watch for signs of these possible complications: hyperglycemia, hypernatremia, hyperkalemia, and fluid imbalance.
- Control infection by maintaining strict aseptic technique during dressing and I.V. tubing changes. Be alert for early signs of local or systemic infection.
- Keep your patient lying in bed during all tubing and dressing changes.

Other Special IVH Nursing Tips

• Infuse IVH solution at a constant rate. If the fluid falls behind schedule, recalculate the time that the infusion will end. *Do not increase the flow rate to catch up:* Radical shifts in fluid rate can cause metabolic problems.

• Generally patients spill +2 urine glucose during the first 2 days of IVH until their pancreas adjusts to the new glucose load.

• If IVH solution is unavailable, give 20% dextrose. Remember, infuse at the same flow rate.

• Discontinue IVH gradually, allowing the body to adjust to lower levels of glucose. Isotonic glucose solution may be administered for at least 12 hours after discontinuation to protect against rebound hypoglycemia.

• Label the IVH solution bottle (date, time of expiration and hanging, rate, and any additives).

• Don't use the IVH line for drawing or giving blood, medications, or piggy-backing other solutions.

• Patients who have a tracheostomy, problems of secretions or a draining wound, or a high-humidity oxygen mask need a waterproof, sterile covering over the IVH site.

• Tell IVH patients that they may have fewer bowel movements.

Ambulatory Hyperalimentation Vest

The ambulatory hyperalimentation vest, illustrated below, allows continuous delivery of parenteral nutrition without restricting the patient's mobility. Made from lightweight, polyester mesh, each vest is adjusted to individual specifications. Its breast pockets accommodate bags of nutrient solution, which are attached to the front of each shoulder by means of swiveled garnet hooks. These pockets vary with the size of the patient and the nutrient bags (for children, the bags are designed to hold 250 ml; for women and small men, 500 ml; for large men, 1,000 ml). Y-tubing connects these bags to a portable volumetric pump, located in a zippered pocket in the vest's lower right quadrant. To provide balance and enhance patient comfort, the pump empties both bags at the same rate.

When the patient is wearing the vest, the 6′ (1.8-m) administration tubing is coiled in one of the pockets. When he's not wearing it, the nutrient bags can hang from a clothes hanger.

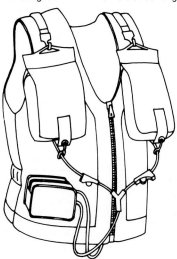

Home I.V. Hyperalimentation

To prepare for home IVH, a barium-impregnated silicone rubber catheter with a Dacron cuff is implanted in the superior vena cava. Its entrance site is on the anterior abdomen, about ¾″ to 1½″ (2 to 4 cm) inferior and lateral to the xiphoid process, to allow the patient to care for the catheter. The catheter's extravascular portion is reinforced with Teflon to reduce the risk of inadvertent catheter fracture. About 2 to 3 weeks after implantation, firm tissue covers the catheter cuff to provide a physical barrier to microbial contamination.

Usually, home IVH patients can ingest part of their caloric requirements, requiring 10 to 14 hours of infusion nightly to supply the remaining nutrients. If all the patient's nutrition must be received intravenously, a continuous infusion may be necessary. For both intermittent and continuous infusion, the patient teaching must include various techniques for proper care.

BASIC PROCEDURES	SPECIAL TOPICS
Catheter heparinization	Detection of complications
Destruction of needles and syringes	Financial support, home referral, medical alert, and networking services
Dressing changes	
Drug administrations	Follow-up appointments
Hand washing	Performance of necessary procedures when traveling
I.V. fat emulsion administration	
I.V. tubing changes	Procurement and storage of supplies
Pump operation	
Self-monitoring (intake and output, daily weight, urine testing, temperature, diet record if patient is allowed partial oral intake)	Schedule for infusion and free time
	Troubleshooting (air embolism; infection; thrombosis; clotted catheter; catheter fracture; pump malfunction; metabolic complications, such as deficiency and excess symptoms; broken I.V. container; contaminated solution or fat emulsion bottle)
Solution preparation	
Use of sterile equipment (packages, gloves, syringes)	
	Weaning

Complications of Intravenous Hyperalimentation (IVH)

CONDITION AND CAUSE	SIGNS AND SYMPTOMS	TREATMENT
Hyperglycemia Too-rapid IVH delivery rate, lowered glucose tolerance, excessive total dextrose load	Glycosuria, nausea, vomiting, diarrhea, confusion, headache, and lethargy; untreated hyperosmolar hyperglycemic dehydration can lead to coma, seizures, and death.	Add insulin to the IVH solution.
Hypoglycemia Excess endogenous insulin production after abrupt termination of IVH solution, or excessive delivery of exogenous insulin	Muscle weakness, anxiety, confusion, restlessness, diaphoresis, vertigo, pallor, tremors, and palpitations	If possible, give carbohydrates orally; infuse dextrose 10% in water, or administer dextrose 50% in water by I.V. bolus.
Fluid deficit Hyperglycemia, vomiting, diarrhea, fistula output, large burns, inadequate fluid replacement, electrolyte imbalance	Fatigue, dry skin and mucous membranes, lengthwise wrinkles in tongue, depressed anterior fontanelle (in infants), tachycardia, tachypnea, decreased urinary output, normal or subnormal temperature, decreased central venous pressure, acute weight loss, hemoconcentration	Increase fluid intake.

Continued

Complications of Intravenous Hyperalimentation (IVH)
Continued

CONDITION AND CAUSE	SIGNS AND SYMPTOMS	TREATMENT
Fluid excess Fluid overload, electrolyte imbalance	Puffy eyelids, peripheral edema, elevated central venous pressure, ascites, acute weight gain, pulmonary edema, pleural effusion, moist crackles	Decrease fluid intake.
Hypokalemia Muscle catabolism, loss of gastric secretions from vomiting or suction, diarrhea; may occur when anabolism is achieved, with its accompanying intracellular movement of potassium	Malaise, lethargy, loss of deep tendon reflexes, muscle cramping, paresthesia, atrial and ventricular dysrhythmias, decreased intensity of heart sounds, weak pulse, hypotension, and complete heart block	Increase potassium intake. Malnourished patient may require an initial dose of 60 to 100 mEq/1,000 calories.
Hypophosphatemia Phosphate deficiency; infusion of glucose causes phosphate ions to shift at start of IVH or within 48 hours of inadequate phosphate intake	Serum PO_4^- levels less than 1 mg/dl can cause lethargy, weakness, paresthesia, glucose intolerance. Severe hypophosphatemia can cause acute hemolytic anemia, convulsions, coma, and death.	Add phosphates to the IVH solution.

Continued

Complications of Intravenous Hyperalimentation (IVH)
Continued

CONDITION AND CAUSE	SIGNS AND SYMPTOMS	TREATMENT
Hypocalcemia Increased doses of phosphates without supplemental calcium; hypoalbuminemia or excess free water	Nausea, vomiting, diarrhea, hyperactive reflexes, tingling of fingertips and mouth, carpopedal spasm, dysrhythmias, tetany, and positive Chvostek's sign and Trousseau's sign	Add calcium to the IVH solution.
Hypomagnesemia Inadequate intake of magnesium; severe diarrhea and vomiting exacerbate hypomagnesemia	Lethargy, tremors, athetoid or choreiform movements, positive Chvostek's sign and Trousseau's sign, painful paresthesia, and tetany	Add magnesium to the IVH solution.
Essential fatty acid deficiency Absent or inadequate fat intake for an extended period	Alopecia, brittle nails, desquamating dermatitis, bruising, reduced prostaglandin synthesis, platelet clumping, thrombocytopenia, enhanced susceptibility to infection, fatty liver infiltration, lipid accumulation in pulmonary macrophages, notching of R waves on EKG, triene to tetraene ratio greater than 0.4.	For the adult patient, infuse two or three bottles of 10% or 20% fat emulsion daily.

Continued

Complications of Intravenous Hyperalimentation (IVH)
Continued

CONDITION AND CAUSE	SIGNS AND SYMPTOMS	TREATMENT
Zinc deficiency Altered requirements associated with stress, the degree of intracellular zinc deficit, and induced zinc deficiencies from redistribution during the anabolism	Diarrhea, apathy, confusion, depression, eczematoid dermatitis (initially in nasolabial and perioral areas), alopecia, decreased libido, hypogonadism, indolent wound healing, acute growth arrest, and hypogeusesthesia (diminished sense of taste)	Add zinc to the IVH solution.
Hypocupremia Long-term administration of IVH, without addition of copper sulfate; infection, high-output enterocutaneous fistulas, and diarrhea predispose to copper deficiency	Neutropenia and hypochromic microcytic anemia	Add copper to the IVH solution.

INDEX

A
ABGs, 24, 28-29, 112-114
Acid-base balance, 27
Adrenal crisis, 132, 135
Adrenal insufficiency, 132
Albumin, 18-19, 20
Aldosterone, 54
Alkaline phosphatase, 20
Amino acids, 51
Anions, 56
AV fistula, 146
AV vein graft, 146

B
Basophils, 12
Bence Jones protein, 51
Bilirubin, 20, 53
Blood pressure, in children, 28-29
Blood studies, routine, 20-23
Blood urea nitrogen (BUN), 4, 20
Bowel obstruction, 151
Burn damage, 168

C
Calcium
  absorption physiology, 82
  blood study, 21
  disorders that affect urine levels, 87
  imbalance, causes of, 88
  importance of, 81
  normal serum and urine values, 81
  phosphorus, and creatinine clearance, 54
  relation to phosphorus levels, 96
  supplements, 86
Capillary membrane, pressure, 99
Capillary pressure effects, 99
Carbon dioxide
  blood study, 21
  tension, 26
Catecholamines, 54
Cations, 56
Chloride, 57, 93, 98
Cholesterol, blood study, 21
Chvostek's sign, 85
Cobalt, 94
Complete blood count (CBC), 10-13
Congestive heart failure, 158
Copper, normal values, 93
Coproporphyrin, in urine, 53
CPR and metabolic alkalosis, 158
Creatinine
  blood supply, 21
  serum, 4, 136
  urine, 51
CSF findings, 32-34, 34-35

D
Dehydration, 105, 106
Dehydration test, 129
Delta-amino levulinic acid, 52
Diabetes insipidus
  dehydration test, 129
  detection, 130-131
  treatment, 128
Diabetic crisis, your role in treatment, 124-125
Diabetic ketoacidosis, 117, 122, 124
Dialysis
  complications, 147
  definition, 143-144
  differences, 141
  indications, 141

INDEX

three forces, 147
Diet, dialysis, 143-144
Diffusion, 147
Diuretics, 108, 110-111
DKA, 118-121, 135
Drug interactions, 48-50, 58-59

E

Edema, 100-101, 102, 169
Elderly, assessment of, 161
Electrolyte, definitions, 24
Electrolyte, imbalances, 135, 155
Electrolyte, serum, functions, 56
Electrolytes, in neuromuscular exchange, 157
Electrolytes, principal extracellular, 55
Electrolyte sources, 57
Endocrine disorders, 134
Eosinophils, 11
Erythrocyte sedimentation rate, 13

F

Femoral vein catheterization, 144-145
Filtration, in dialysis, 147
Fluid, assessment of, 106
Fluid balance, 2-3
Fluid concentration, 44
Fluid deficit, 180
Fluid excess, 103, 181
Fluid imbalance, 156
Fluid loss, internal, 150
Fluid replacement
  guidelines, 170-171
  how to assess, 153
  types, 172-173
Foley catheter, 42

Food, high in potassium, 65
Food, low in sodium, 65
Fractures, 143

G

Gastric tubes, 88
Gastrointestinal bleeding, 86
Glomerular filtration rate, 139
Glucose (fasting), 22

H

Harris tube, 89
Heart failure, symptoms of, 154
Hematocrit, 4, 13
Hemodialysis, 141
Hemodialysis access, 144-146
Hemodialysis complications, 148
Hemoglobin, 12, 52
Hepatojugular reflux, 155
HHNC, 118-121, 135
Home I.V. hyperalimentation, 179
Homeostasis
  absorption and excretion, 1
  hormonal secretion sites, 8
  hormones, 7
  role of various organs, 6
Hormones, and hypoglycemia, 127
Hypercalcemia, 84, 88
Hyperglycemia, 180
Hyperkalemia, 71, 74
Hypermagnesemia, 91, 182
Hypernatremia, 63
Hyperphosphatemia, 97
Hypocalcemia, 83, 88
Hypoglycemia, 118-121, 126, 127, 180
Hypokalemia, 70, 181
Hypomagnesemia, 90

INDEX

Hyponatremia, 62
Hypoparathyroidism
  acute pathophysiology, 133
  resulting electrolyte imbalances, 135
Hypophosphatemia, 96, 181
Hypovolemia, 152
Hypovolemic shock, 149

I

Intravenous hyperalimentation (IVH), 175-176, 177, 180-183
Iron, 57
I.V.s, 162-167

L

Laboratory tests for evaluating fluid status, 4-5
Laboratory values, how to interpret, 26-27
Lab values in pancreatitis, 174
Lactic dehydrogenase, normal range, 22
Lymphocytes, normal range, 10

M

Magnesium
  importance, 89
  normal serum and urine values, 39
  supplements, 92
Melanin, in urine, 53
Metabolic acidosis, 112-113, 116
Metabolic alkalosis, 112-113, 115, 158
Monocytes, 12
Myocardial infarction, 28
Myoglobin, in urine, 52
Myxedema crisis, 135

N

Neuromuscular functioning, role of electrolytes, 157
Neutrophils, 11

O

Osmolarity, 9
Osmosis, 147
Overhydration, 104
Oxygen tension, 26

P

Pancreatitis, 174
Peritoneal dialysis
  acute care, 142
  compared to hemodialysis, 141
  how it works, 142
Peritoneal fluid, analysis, 36
Peritoneal fluid, causes of abnormalities, 37
Phosphate
  disorders that affect urine levels, 87
  normal serum values, 93
Phosphorus, source of, 57
Pinch test, 107
Plasma, 16-17
Plasma protein fraction, 18-19
Plasma protein incompatibility, 177
Platelets, 13, 16-17
Pleural fluid, 30-31
Porphobilinogen, in urine, 52
Potassium
  blood supply, 22
  dietary sources, 79-80
  electrolyte sources, 57
  imbalances, 72-73
  importance, 69

INDEX

in DKA, 123
normal serum values, 69
supplements, 75-78
Protein, total, blood study, 23
Proteins, in urine, 51
Psoas sign, 84
Pulmonary edema, 48
Pulse, in children, 28-29

R

Red blood cells, 12, 14-15
Renal deterioration, 139
Renal failure, 137, 138, 140
Renal system, 137
Respirations, in children, 28-29
Respiratory acidosis, 112-113
Respiratory alkalosis, 112-113
Respiratory exchange rate, 25
Right heart failure, 34
Rovsing's sign, 84

S

Salt, how to avoid, 64
Serum glutamic-oxaloacetic transaminase (SGOT), 23
Serum osmolality, 4
17-hydroxycorticosteroids, 54
17-ketosteroids, in urine specimen, 54
Skin grafts, temporary, 124
Sodium, 22, 57, 60, 68
Sodium and aldosterone, 61
Sodium pump, 64
Sodium supplements, 66-67
Subclavian vein catheterization, 145-146
Surgery, fluids critical in, 160-161
Surgery, pre- and post- considerations, 159
Synovial fluid, 38-39, 40

T

Temperature, in children, 28-29
Thiazide diuretics, 109
Thyroid crisis, 132
Thyrotoxic crisis, 135
Tissue perfusion, 55
Trace elements, 94
Trousseau's sign, 85

U

Urea, in urine, 51
Urea nitrogen, blood values, 136
Uric acid, in urine, 23, 52, 136
Urinary output, 44
Urine, 24-hour specimen, 54
Urine color, 45
Urine formation, 41-43
Urine osmolality, 5
Urine pH, 5
Urine specific gravity, 5
Urine specimen analysis, 46-47, 51-53
Urobilinogen, in urine, 53
Uroporphyrin, in urine, 53

V

Venipunctures, 23
Vital signs, 28-29

W

WBC differential, 10-12
Weight, accuracy, 3
White blood cells, 10, 14-15
Whole blood and components, 14-19
Wilson's disease, 95

Save 36% on the world's most trusted nursing journal, *Nursing86*®.

More nurses depend on *Nursing86* than any other professional journal. Subscribe now and get over one third off the price of a full year's subscription—12 issues in all. There's no risk: You can stop your subscription at any time and get a complete refund on all unmailed issues.

Send no money now.
We'll bill you later for $17.95. Or include your payment with your order. Mail your name and address to:

> ***Nursing86* Savings Offer**
> One Health Care Drive
> P.O. Box 2021
> Marion, OH 43305

A tax-deductible professional expense for nurses working in the U.S.
Basic annual rate is $28.
© 1985 Springhouse Corporation

Nursing86
THE WORLD'S LARGEST NURSING JOURNAL
APRIL

Beware the red menace! Eliminate decubitus ulcers

Anticonvulsants

The act of touching so simple so soothing

Help your cancer patient find *his* style of coping

R.E.A.C.T.—a new way to measure